Squirrels: *The Animal Answer Guide*

Squirrels

The Animal Answer Guide

Richard W. Thorington Jr. and Katie Ferrell

The Johns Hopkins University Press   Baltimore

The Johns Hopkins University Press
2715 North Charles Street
Baltimore, Maryland 21218-4363
www.press.jhu.edu

Library of Congress Cataloging-in-Publication Data

Thorington, Richard W.
    Squirrels : the animal answer guide / Richard W. Thorington, Jr., and
Katie E. Ferrell.
        p. cm
    Includes bibliographical references and index.
    ISBN 0-8018-8402-0 (hardcover : alk. paper) — ISBN 0-8018-8403-9
(pbk. : alk. paper)
    1. Squirrels. I. Ferrell, Katherine E., 1974– II. Title.
QL737.R68T48 2006
599.36—dc22   2005032061

A squirrel leaping from bough to bough, and making the wood but one wide tree for his pleasure, fills the eye not less than a lion,—is beautiful, self-sufficing, and stands then and there for nature.

RALPH WALDO EMERSON

Contents

Acknowledgments xi
Introduction xiii

1 Introducing Squirrels 1
What are squirrels? 1
How many kinds of squirrels are there? 7
Where do squirrels live? 8
Why are there no flying or ground squirrels in
    South America? 10
How are squirrels classified? 10
What is the current classification of squirrels? 13
What characterizes the major groups of squirrels? 15
When did squirrels evolve? 23
What is the oldest fossil squirrel? 23

2 Form and Function 25
What are the largest and smallest squirrels? 25
How fast does a squirrel's heart beat? 25
Can squirrels see color? 26
Do all squirrels have cheek pouches? 27
Can squirrels swim? 28
How far can squirrels jump? 29
Do squirrels fly? 31
What are the largest and smallest flying squirrels? 31
How do you make a flying squirrel? 32
How far can flying squirrels glide? 35
How can you determine whether a fossil squirrel was a tree
    squirrel, a flying squirrel, or a ground squirrel? 37

3 Coat Color and Squirrel Genetics 38
What species are the black squirrels I see? 38
What species is the red squirrel? 40
What causes the different coat colors of squirrels? 41
How are hair colors determined genetically? 41
What about patterns of coat color? 43
Are there age-related differences in coat color? 45

Are there seasonal differences in coat color? 45
Is there much geographic variation in squirrel species? 46

4 Squirrel Behavior 48
Are squirrels social? 48
Do squirrels fight? 52
How smart are squirrels? 52
Do squirrels play? 55
Do squirrels talk? 56
How do squirrels avoid predators? 58

5 Squirrel Ecology 69
Where do squirrels sleep? 69
Do squirrels migrate? 71
How many squirrel species coexist in a forest? 72
Why are there so many more species of squirrels in African
      and Asian forests? 74
How do squirrels survive in the desert? 76
How do squirrels survive the winter? 79
What is hibernation? 82
Do squirrels have enemies? 83
Why do squirrels commit infanticide? 84
Do squirrels get sick? 86
How do squirrels influence vegetation? 88

6 Reproduction and Development 92
How do squirrels reproduce? 92
How long are female squirrels pregnant? 95
Where do mother squirrels give birth? 95
How many babies do squirrels have? 96
Are all littermates equally related? 97
How long do female squirrels nurse their young? 98
How fast do squirrels grow? 98
How long do squirrels live? 100

7 Foods and Feeding 102
What do squirrels eat? 102
How do squirrels open hard nuts? 106
Do all squirrels bury their food? 107
How does the squirrel decide what food to store? 110
How do squirrels find the food they have buried? 112

8   Squirrels and Humans 114

Do squirrels make good pets? 114

Should people feed squirrels? 114

Do squirrels feel pain? 115

What do I do if I find an injured or orphaned squirrel? 115

How can I become a better observer of squirrels? 116

How do I know whether I have flying squirrels in
    my backyard? 118

Why are squirrels important? 119

9   Squirrel Problems (from a human viewpoint) 124

Are squirrels pests? 124

How do I keep squirrels away from my . . . ? 126

Are squirrels vectors of human disease? 130

10   Human Problems (from a squirrel's viewpoint) 131

Are squirrels endangered? 131

Will squirrels be affected by global warming? 134

Are squirrels ever invasive species? 137

Do people hunt and eat squirrels? 138

Are squirrel-hair brushes actually made of squirrel hair? 139

Why do so many squirrels get hit by cars? 140

11   Squirrels in Stories and Literature 141

What roles do squirrels play in religion and mythology? 141

What do squirrels have to do with the Cinderella story? 143

What roles do squirrels play in popular culture? 144

How are squirrels incorporated into poetry? 144

How are squirrels incorporated into literature? 148

12   "Squirrelology" 152

Which species are best known? 152

Which species are least known? 154

How do scientists recognize individual squirrels? 156

Appendix: Squirrels of the World 157

Bibliography 165

Index 179

Acknowledgments

Many persons have shared and encouraged our enthusiasm for the study of squirrels. We would like to thank them all, but we must restrict our acknowledgments to those who have played special roles. These include Thorington's mentors in graduate school, Charles Lyman and Ernst Mayr, who directed him into the field of mammalogy, and Ferrell's mentor, Gerald Svendsen, who gave her an appreciation for mammalian biology. Over the years, Thorington has collaborated with and learned from a number of colleagues and assistants, including C. Gregory Anderson, Amy Betts, Andrea Cardini, Jeff Chu, Rich Cifelli, Lindsay Pappas Davis, Karrie Darrow, John Eisenberg, Robert Emry, William Glanz, Lawrence Heaney, Robert Hoffmann, Sharon Jansa, Jennifer Leonard, Jesus Maldonado, Amy Musante, Diane Pitassy, Louise Roth, Chad Schennum, Brian Stafford, and Robert Voss, to all of whom he is grateful.

At the Smithsonian, we have been privileged to work with a large museum collection of squirrels, perhaps the largest in the world, and a superlative library. In addition, we have had financial support from the Smithsonian, not just with the preparation of the book but also for fieldwork, visits to other museums, and research on the collections. The Environmental Science Program and Scholarly Studies Program have supported much of Thorington's work. Thorington is especially thankful for the institutional support that he has received, enabling him to continue working although quadriplegic.

The Division of Mammals has been Thorington's intellectual home for 36 years, and he wishes to express his appreciation to past and present colleagues there for many informative discussions and debates about mammalian biology, especially taxonomy and systematics. Don Wilson suggested this book and encouraged us to write it, and Bob Hoffmann has been a wonderful source of information and enthusiasm. In the writing of this book, we have been guided and encouraged by Vincent J. Burke, Senior Editor, of the Johns Hopkins University Press, who responded quickly to all our questions and concerns. We would also like to thank Michael Steele for his thoughtful comments and suggestions during the review process. We are grateful for poetry and literature quotes on squirrels from Phil Clapham and the members of "QDJ," especially Julie Warfield, Sharon Young, and Yulia Alper.

Finally, we are delighted to acknowledge the shared enthusiasm of our respective families for natural history and their willingness to indulge our fascination with squirrels.

Introduction

In the late afternoon, Thorington and his host drove to a shrine on the outskirts of Tokyo. They walked up a broad, well-maintained path from the parking lot into the shrine. Above them were magnificent old Japanese cedars, the famed cryptomeria trees; the understory consisted of maples and a variety of other trees that Thorington did not recognize. His host, Dr. Imaizumi, pointed to a tree hole, and Thorington trained his binoculars on it. Soon, a squirrel's head appeared, giving him his first glimpse of a wild Japanese giant flying squirrel. He watched impatiently as the squirrel withdrew its head, then reappeared, and again withdrew its head. Suddenly, it emerged and seemed instantly to dash to the backside of the tree and out of sight. He vividly remembers his sinking feeling: he had seen a giant flying squirrel but just barely. Then, from higher in the tree, the same squirrel appeared and launched out into a gliding flight to the base of another tree. Soon the darkening shrine seemed full of squirrels, calling excitedly and gliding back and forth overhead. They had been confined to their nests for several days by a typhoon, and this was their first dry, windless night. Too soon, it was dark and Thorington followed Dr. Imaizumi as they retraced their steps to the car, but the thrill of that evening, 30 years ago, is still with him.

Thorington's interest in squirrels started long before then, probably with his unsuccessful attempts as a boy to keep them from raiding his bird feeder. When he came to the Smithsonian 36 years ago, his enthusiasm increased when he began to examine the worldwide diversity in the museum collection, from giant flying squirrels to pygmy tree squirrels, from large Asian marmots to least chipmunks, from common eastern gray squirrels to rare Borneo sculptor squirrels. It always helps to have someone to share your enthusiasm with, and he was fortunate to find an extremely enthusiastic high school student, Larry Heaney, already at work helping to curate the squirrel collection. Larry's cry, "Sciuridae Forever!" still seems to resonate through the hallways of the Smithsonian, although now he is the much more sedate, silver-haired Curator of Mammals at the Field Museum in Chicago. Throughout their careers Thorington and Heaney have continued to collaborate on squirrels.

As a curator of one of the largest squirrel collections in the world, Thorington answers many questions about squirrels, asked by a wide variety of people. This book is designed to provide answers, both to these questions and to questions that we think should be asked. As we have reviewed the

Inquisitive, *industrious*, *talkative*, and *energetic* are terms commonly used to describe the squirrel. Photo © Plaistow John, www.scarysquirrel.org

Many people may not realize that prairie dogs, like this black-tailed prairie dog (*Cynomys ludovicianus*) mother and pup, are also squirrels. In all, there are 278 species of squirrels around the world. Photo © Shirley Curtis, www.scarysquirrel.org

literature on squirrel biology, we have looked for facts and interrelationships of general interest, and these we have arranged as answers to relevant questions.

One of the exciting things about focusing on a group of animals is that when you look closely, fascinating connections appear. For example, there are interrelationships between the fossil history of the squirrels and their current distribution around the world; their anatomy and their ecology; and their ecology and their behavior.

The worldwide diversity of squirrels is astounding. There are 278 species, which occur on five continents. We like to say, "The sun never sets on the Sciuridae!" This diversity encourages informative comparisons. The

Introduction

ground squirrels and tree squirrels of Africa exhibit differences from those of North America and northern Eurasia, which are different again from those of Southeast Asia.

These differences raise tantalizing questions about their behavior, ecology, morphology, and evolutionary history. Some of these questions we can answer, and some of them we cannot. Therefore, an animal answer guide, like this one, is inevitably a work in progress. We hope our readers will view it as such, and that our answers will provoke more questions and stimulate more inquiries into the biology of squirrels.

Squirrels: *The Animal Answer Guide*

Chapter 1

Introducing Squirrels

What are squirrels?

Squirrels are among the most widely known and recognized mammals, inhabiting all continents except Antarctica and Australia. In many parts of the world they occupy human habitats, gladly sharing our lunches in a city park, helping to empty our bird feeders, or feeding on our crops. Squirrels belong to a family of rodents, the Sciuridae, whose common ancestor lived some 30 to 40 million years ago and gave rise to the 278 species we currently assign to the family. All squirrels share a number of anatomical features—teeth, jaw musculature, skull, and other bones—that scientists use to identify recent and fossil squirrels. There are also important differences in anatomy between the main groups of squirrels. Below we discuss some basic components of squirrel anatomy, including similarities and differences between tree squirrels, flying squirrels, and ground squirrels.

TEETH AND SKULL. Squirrels, like all rodents, have four chisel-like incisors at the front of the mouth—two above and two below—which they use for gnawing. These incisors grow continuously and have roots that extend well back into the maxilla and mandible, unlike human teeth, which have short roots and grow only so long. Constant gnawing keeps rodent incisors short and sharp, because the enameled layer on the outside of the tooth wears much more slowly than the inner layers of dentine. Squirrels use their incisors to gnaw almost anything, but under natural conditions they use them most commonly to cut vegetation and to gnaw hard objects like nuts. If for some reason a squirrel cannot gnaw, its incisors will continue to grow until the animal dies because it can no longer eat.

Eastern gray squirrels (*Sciurus carolinensis*) are perhaps the most widely recognized squirrel species in the world. Photo © Donald Reeve, www.scarysquirrel.org

Cheek teeth are separated from the incisors by a space called the diastema. A squirrel's cheek teeth consist of one or two premolars and three molars in the upper jaw, and one premolar and three molars in the lower jaw. These have short roots and determinate growth, meaning they grow for a short while and then stop. Squirrels use their molars and premolars for grinding up food before they swallow it, just as you do. The molars and premolars have cusps (called cones on the upper teeth and conids on the lower teeth) connected by ridges (called lophs on the upper teeth and lophids on the lower teeth). The differences and similarities in patterns of these cusps and ridges on teeth are used extensively by paleontologists to identify fossils and to recognize which fossil mammals are closely related to which others.

A squirrel must be able to move its jaw forward to bring its incisors together to gnaw, but it also must be able to move its jaw backward to bring the molars together to grind. To move the jaw forward to gnaw, some of the jaw muscles must attach farther forward on the skull than they do on the lower jaw. Different kinds of rodents have evolved jaw muscles that attach farther forward on the skull in several different ways. In all but the most ancient fossil squirrels, these muscles attach below and in front of the orbit of the eye, and also (in most squirrels) on the side of the nose. This type of jaw muscle attachment is called sciuromorphy. It is also found in a handful of other rodents, like beavers, pocket gophers, and kangaroo rats.

HANDS AND FEET. The hands and feet of squirrels are relatively unspecialized, with five digits on each, although the thumb ranges from small to diminutive. Like you, they have three phalanges, or bones, on each finger

Squirrels: The Animal Answer Guide

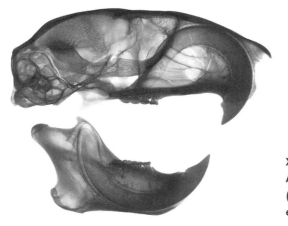

X-ray of the skull of the southern Asian tufted ground squirrel (*Rheithrosciurus macrotis*), with its ever-growing incisors.

and toe, except for the thumb and the first toe, each of which has two. The relative proportions of the phalanges of squirrels, particularly of ground squirrels, are different from yours, and they have claws instead of nails on the terminal phalanges of all digits except for the thumb. There is a small nail on the thumb, but it may not be large enough to be seen. The hands of burrowing squirrels tend to be wider than those of tree squirrels, with the middle digit the longest, presumably because a wider hand is more useful for digging. In tree squirrels, the fourth digit is the longest and serves as an effective grappling iron when the squirrel climbs a vertical trunk. The hands and wrists of flying squirrels also have an extra addition, the styliform cartilage, which extends laterally from the heel of the hand and assists in the support of the gliding membrane.

One of the most striking features of tree and flying squirrels is their ability to turn their hind feet around when they are coming down a tree head-first. For you, the equivalent move would be to stand on your tiptoes, rotate your ankles so that the soles of your feet face each other, then keep rotating them until the soles of your feet point forward—without moving the rest of your leg! Most squirrels accomplish this with the extraordinary mobility of the joint between the most proximal ankle bone, the astragalus or talus, and the heel bone, the calcaneus. The ankle anatomy of some squirrels (for example, the Holarctic tree squirrels: those found in the northern parts of North America and Eurasia) shows certain specializations for this move, whereas in other squirrels (the Southeast Asian tree squirrels) the same specializations are not present. However, these squirrels still are able to turn their feet around. We just aren't yet sure how they do it.

Another notable difference in squirrel anatomy is limb length. Ground squirrels usually have shorter limbs than tree squirrels for their body size. Tree squirrels need long, muscular legs for leaping from branch to branch and long arms to reach around a tree trunk. By contrast, a burrow-dwelling

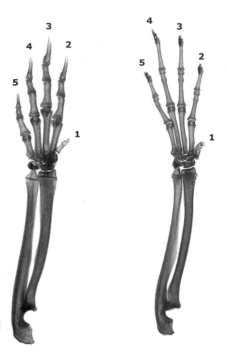

X-ray of the forearm of a ground squirrel and a tree squirrel of comparable size. On the left is a Richardson's ground squirrel (*Spermophilus richardsonii*); on the right is the North American red squirrel (*Tamiasciurus hudsonicus*). Notice how digit no. 4 is much longer on the tree squirrel than on the ground squirrel.

ground squirrel has short, stout limbs well suited for digging. Flying squirrels have the longest limbs relative to body size of all the squirrels. This appears to be a necessary part of the flying squirrel adaptation. (For more information on flying squirrel anatomy see "How do you make a flying squirrel?" in chapter 2.)

Some Holarctic ground squirrels have another specialization, not seen in tree squirrels: an extra layer of the deltoid muscle of the shoulder that extends to the forearm. This addition assists in flexing the forearm and is presumably important in digging. It is not found in chipmunks (*Tamias*) or antelope ground squirrels (*Ammospermophilus*), so it must have evolved among the other ground squirrels (*Spermophilus*, *Marmota*, and *Cynomys*) after the chipmunks and antelope ground squirrels diverged from the lineage. It also is not found in any of the African or southern Asian ground squirrels.

TAILS. Many ground squirrels have relatively short tails, which are probably useful in a burrow, particularly for feeling their way when backing up. Tree squirrels, on the other hand, have longer tails, which are useful for balance when leaping and running across narrow branches. All squirrel tails are most flexible at the base of the tail. This is because they have many short vertebrae, thus more intervertebral joints, at the base of the tail, and longer vertebrae toward the middle and end of the tail. The characteristic way a tree squirrel holds its tail, curved up against the back, demonstrates this attribute quite nicely. The hairs on the tails of most tree squirrels are haired

Squirrels: The Animal Answer Guide

An eastern gray squirrel outside Washington, D.C., pauses while descending a tree headfirst—a feat made possible by its specialized ankle anatomy. Photo © Caroline Thorington

distichously, meaning the hairs along the sides are much longer than the hairs along the top or bottom. Among the giant tree squirrels, the hairs on the tail are instead approximately the same length in every direction. We think this reflects the effective wind resistance of a broad tail, thus assisting with balance in a small squirrel (perhaps up to 1 kg, or 2.2 lbs), contrasting with the need for a longer, more massive tail in a large squirrel (more than 1.5 kg, or 3.3 lbs). The longer the hairs, the more the hairs bend and the less wind resistant they are. A large squirrel, like a giant tree squirrel, cannot grow long enough, thick enough hair to be wind resistant and thus effective for assisting balance.

Another important function of tails is in heat conservation or dissipation. Long cylindrical objects, like rodent tails, are difficult to insulate and they lose heat rapidly because they have so much surface area per unit of length. Although this is convenient if the animal is overheated, it is inconvenient if it is trying to keep warm. Squirrels, like other rodents, have a complex vascular system in their tails—a "countercurrent heat exchange system" within a vascular bundle on the ventral side of the base of the tail. In cold conditions, the warm blood entering the tail by the large ventral arteries warms the venous blood returning to the body in the adjoining ventral veins, and in so doing, the arterial blood cools. In other words, heat

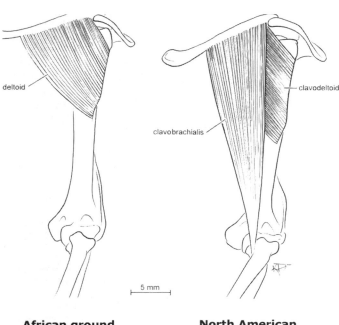

deltoid

clavodeltoid

clavobrachialis

5 mm

The clavobrachialis muscle, found in the Holarctic ground squirrels (tribe Marmotini) but not in African ground squirrels (tribe Xerini).

Photo © Karolyn Darrow

African ground squirrel
Xerus

North American ground squirrel
Spermophilus

is returned to the body via the venous blood, not lost to the environment. In warm conditions or when the animal is overheated, much of the venous blood is returned to the body through large lateral veins. This blood is not warmed in the countercurrent system, and so the arterial blood to the tail is not cooled. The warm arterial blood warms the skin of the tail, which then loses heat to the environment. The result is that under cold conditions, the tail itself is not warmed up, and little body heat is lost. But under warm conditions, the tail is warmed and heat is lost from the whole surface to the air.

The tail is also used in thermoregulation by serving as a blanket over the squirrel in cold weather and, among some squirrels, as a parasol to protect them from the heat of the sun in hot weather.

VIBRISSAE. Hairs are very sensitive to touch, as everyone knows, and squirrels have long, stiff hairs, called vibrissae, dedicated to the detection of tactile stimuli. There are a dozen to two dozen vibrissae on the side of the nose above the mouth and three to six vibrissae on the side of the cheek. There are usually four vibrissae above the eye, rooted at the front edge, several on the underside of the chin, and several farther back but above the throat. There are also vibrissae near the wrist, at the lower end of the forearm just above the lateral side of the wrist. All whiskers are vibrissae, but not all vibrissae are whiskers.

Squirrels: The Animal Answer Guide

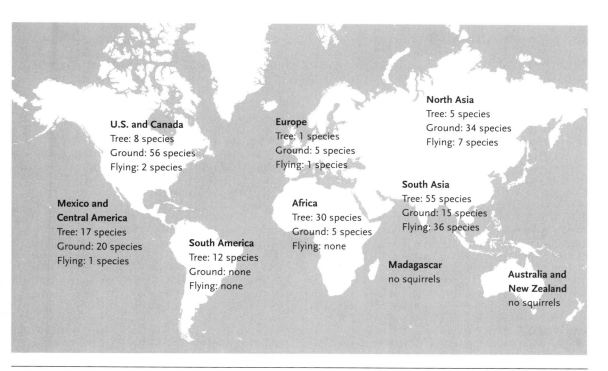

U.S. and Canada
Tree: 8 species
Ground: 56 species
Flying: 2 species

Europe
Tree: 1 species
Ground: 5 species
Flying: 1 species

North Asia
Tree: 5 species
Ground: 34 species
Flying: 7 species

Mexico and Central America
Tree: 17 species
Ground: 20 species
Flying: 1 species

South America
Tree: 12 species
Ground: none
Flying: none

Africa
Tree: 30 species
Ground: 5 species
Flying: none

South Asia
Tree: 55 species
Ground: 15 species
Flying: 36 species

Madagascar
no squirrels

Australia and New Zealand
no squirrels

A world map showing the distribution of squirrels. The distinction between tree squirrels and ground squirrels is arbitrary in some cases.

How many kinds of squirrels are there?

We have recently estimated that there are 278 species, or unique types, of squirrels. This number is likely to increase as we continue to study squirrels more carefully, in particular, as we learn more about their DNA. The ecological diversity of squirrels is also large. In North America, there are tree squirrels, flying squirrels, ground squirrels, chipmunks, marmots, prairie dogs, and antelope ground squirrels (so named because they are speedy little fellows and show pale rumps when they bound away). In South America there are tree squirrels and pygmy tree squirrels, but no flying squirrels and no ground squirrels. The lack of diversity in South America is not surprising because squirrels have been on that continent for only a few million years. In Africa, where squirrels have been for 20 million years, there is a greater diversity of squirrels. There are tree squirrels, a pygmy tree squirrel, ground squirrels, and bush squirrels, but no flying squirrels. (There are scaly-tailed flying squirrels in Africa, but they are a completely different group of rodents from the true squirrels.) In Eurasia, there are tree squirrels, marmots, and chipmunks—all closely related to the North American ones—and a few species of flying squirrels. In Southeast Asia there are the beautiful tree squirrels, the giant tree squirrels, four species of pygmy squir-

A marmot, better known for its ground-dwelling habits, demonstrates its talent for tree climbing.

rels, a large number of flying squirrels, and several more specialized squirrels. We frequently refer to Southeast Asia as the "squirrel headquarters of the world" because of the extraordinary diversity there.

Where do squirrels live?

Squirrels are found in an extraordinarily diverse range of habitats, from desert conditions to extremely wet rain forests. The greatest diversity of squirrel species is found in warmer climates, specifically the tropics, but species of ground squirrels and marmots occur in the Arctic, where they hibernate through the long, cold winter. Squirrels occur naturally on all continents except Antarctica and Australia. As far as we know, squirrels never have reached Madagascar, Greenland, or any of the truly oceanic islands, like Hawaii or the Galápagos Islands. However, squirrels are found on the Philippine Islands and Sulawesi.

We commonly divide the squirrels into three groups: the tree squirrels, the ground squirrels, and the flying squirrels. For the most part, this division is a good description of the animals' habits. Tree squirrels and flying squirrels usually live and nest in the trees, sometimes in tree hollows and at other times in leaf nests outside of hollows. Many of them, however, forage

Squirrels: The Animal Answer Guide

A rock squirrel (*Spermophilus var-iegatus*) at Grand Canyon National Park in Arizona. Photo © Tim Coffer

for food on the ground, like the eastern gray squirrel (*Sciurus carolinensis*) and the northern flying squirrel (*Glaucomys sabrinus*), and some of them, like the North American red squirrel (*Tamiasciurus hudsonicus*), may even nest underground. Ground squirrels usually nest in underground burrows and forage on the ground. Chipmunks (like the eastern chipmunk, *Tamias striatus*, and the Townsend's chipmunk, *Tamias townsendii*, in the West) and even marmots (like the woodchuck, *Marmota monax*) will climb trees, and some chipmunks will nest in trees, for example, the shadow chipmunk (*Tamias senex*).

Outside North America, squirrels show even greater diversity in their habits. Some members of the Southeast Asian tree squirrel group are strictly arboreal, while others are strictly terrestrial. The woolly flying squirrel (*Eupetaurus cinereus*) and the complex-toothed flying squirrel (*Trogopterus xanthipes*) spend the daytime sleeping in caves, not trees, and the Asian long-clawed ground squirrel (*Spermophilopsis leptodactylus*) sometimes burrows in sand dunes. The Himalayan marmot (*Marmota himalaya*) inhabits parts of the Himalayan Mountains of Nepal, Tibet, and India at high elevations from 4,000 meters to 5,500 meters (13,123 to 18,035 feet) and is considered one of the highest-living mammals in the world.

There is a well-known "Harvard Law of Animal Behavior" which states, "Under the best controlled experimental conditions, the animal will do

Lush vegetation provides food and shelter for eastern gray squirrels in the tropical habitats of southern Florida. Photo © Donald Reeve, www.scarysquirrel.org

whatever it darn well pleases." Squirrels in the wild seem well acquainted with this law! They opportunistically exploit their environment, in the trees, on the ground, and under the ground, irrespective of how we categorize them.

Why are there no flying or ground squirrels in South America?

With the possible exception of the pygmy squirrel (*Sciurillus*), squirrels did not get to South America until relatively recently, in geological terms, after the Panama land bridge formed between North and South America. At that time, there most likely were no flying squirrels or ground squirrels in southern Central America, just like today. Consequently, since only tree squirrels were present in the southernmost part of Central America, it could only be tree squirrels that would cross the Panama land bridge into South America, which probably occurred within the last 3.5 million years. At present, the southern flying squirrel (*Glaucomys volans*) has a range only as far south as Honduras, and the southernmost ground squirrels (*Spermophilus*) reach no farther south than Mexico.

How are squirrels classified?

Classification is a way to organize large amounts of information. To classify life on earth, scientists use a hierarchical classification system that starts broadly and becomes more specific. Organisms, on the basis of similarities or differences, are placed first into kingdoms, which are subdivided

Squirrels: The Animal Answer Guide

Grasslands are the habitat of choice for Richardson's ground squirrels (*Spermophilus richarsonii*) in southwest Montana. Photo © Caroline Thorington

into phyla or divisions. These in turn are divided into classes, then into orders, then families, and finally into genera and species. Every known organism is identified by two Latin terms: the genus name and the species name. These two names together are the official scientific name of the species. For example, the scientific name for the eastern gray squirrel is *Sciurus carolinensis*. (Notice that the name of the genus is capitalized, but the name of the species is not.) This system, called binomial nomenclature, was developed by biologist Carolus Linnaeus in 1758. Other systems of classification and nomenclature are being debated, but they will not be treated in this book, with the exception of cladistics, which emphasizes groupings (clades) of organisms sharing a common ancestor and includes that ancestor and all its descendents in the clade.

Squirrels, like all organisms, are classified by the way they are related to one another. All squirrels are more closely related to one another than they are to any other mammals, and so we put them together into a family, the clade of all squirrels. Similarly, families can be divided into subfamilies that contain squirrels that are more closely related to each other than they are to any other squirrels. The same is true for every tribe, genus, and so on. The trick is to determine how different species and groups of species are related. We determine these relationships in two main ways: by looking at

Introducing Squirrels

similarities and differences in morphology and in DNA. With correct interpretation of which features are ancestral and which features are derived, we can determine the genealogical relationships, or the phylogeny, among squirrels. These relationships are commonly represented as cladograms, which are stick-figure trees showing how the species and clades of species are related.

DNA is a particularly useful tool for determining phylogenies, because there are so many parts of the molecule (specifically, base pairs) subject to change. The DNA of an organism is replicated every generation, so every squirrel has DNA almost exactly like that of its father and mother. Small mistakes in the replication do occur, and these mistakes are passed on to the subsequent generation when the squirrel breeds. We like to use the analogy of medieval scribes, who, in the time before printing presses, copied manuscripts by hand. If a scribe made a mistake in copying, the subsequent copies of a manuscript made from his original would incorporate the same mistake. Subsequent scribes might make additional mistakes not found in the one from which they copied but incorporated into later copies. Thus, historians can trace the origins of differing manuscripts from the mistakes that have or have not been incorporated into them. Similarly, we can determine the evolution of squirrels from the errors of DNA replication that have or have not been passed on to subsequent generations.

The process is actually more complicated than we have portrayed it. DNA degrades over time, so all the DNA used in our molecular research is from modern squirrels. This means that the squirrel family tree must be constructed by working backward from the tips of the branches. If an animal or a group of animals, like the giant tree squirrels, does not have any close relatives, then it will be represented as the tip of a long branch. Long branches can be problematic, because they can conceal repeated replication errors. To return to our analogy of medieval scribes, it is as if a series of scribes wrote "the" instead of "two" and "two" instead of "the." After a series of such replication errors, these words would be almost useless for interpreting the lineage of the resulting manuscript. Computer programs are used to assess the probabilities that such problems occur and the likelihood that a particular tree is the best estimate of the true tree.

Some genes evolve more quickly than others and are better for estimating recent divergences between closely related species. Mitochondrial genes are commonly used for this purpose. These genes are found in the mitochondria of the cell and are inherited solely from the mother. Other more slowly evolving genes, usually nuclear genes, are better for resolving more ancient divergences. The nuclear genes are found on the chromosomes within the nucleus and are inherited from both parents. We will not go into further description of the techniques used for analyzing DNA. The

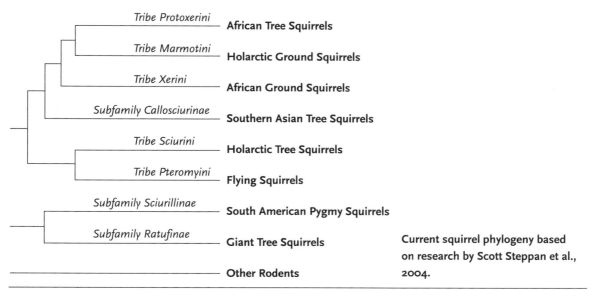

Tribe Protoxerini	**African Tree Squirrels**
Tribe Marmotini	**Holarctic Ground Squirrels**
Tribe Xerini	**African Ground Squirrels**
Subfamily Callosciurinae	**Southern Asian Tree Squirrels**
Tribe Sciurini	**Holarctic Tree Squirrels**
Tribe Pteromyini	**Flying Squirrels**
Subfamily Sciurillinae	**South American Pygmy Squirrels**
Subfamily Ratufinae	**Giant Tree Squirrels**
	Other Rodents

Current squirrel phylogeny based on research by Scott Steppan et al., 2004.

analytic techniques themselves are evolving rapidly, and there is much experimentation and ongoing discussion about which methods are best.

What is the current classification of squirrels?

Two recent studies of the DNA of squirrels, one by John Mercer and Louise Roth at Duke University and a second by Scott Steppan and his colleagues at Florida State University, have greatly increased our knowledge of squirrel phylogeny. Their conclusions are in good general agreement with one another and with previous phylogenies based solely on anatomical features, but there are some differences.

Squirrels belong to the order Rodentia, and within that belong to the family Sciuridae. Based on the current morphological and molecular studies, the family Sciuridae is presently divided into five subfamilies: Sciurillinae, Ratufinae, Sciurinae, Callosciurinae, and Xerinae.

Squirrels evolved as a distinct lineage of rodents approximately 36 million years ago, at the end of the Eocene or beginning of the Oligocene geological epic. In both of the DNA studies, the most ancient squirrels are the pygmy tree squirrels of South America (Sciurillinae) and the giant tree squirrels of southern Asia (Ratufinae). Sciurillinae contains only a single recognized species, but we believe that further study will prove this single species actually is a collection of several closely related species. Ratufinae contains the four species of giant tree squirrels of Southeast Asia. Among these, the Indian giant squirrel is one of the most beautiful squirrels in the world, with dark maroon and black hair on the back and tail, cream-colored legs, feet, and tail tip, and maroon ear tufts. In the Mercer and Roth molecular study, the Sciurillinae and Ratufinae seem to have branched off independently, the pygmy squirrels diverging first. Scott Steppan and his colleagues contested this with their data, which suggest that the two subfamilies diverged from a common branch of the squirrel tree. It is a bizarre idea that the South American pygmy squirrel and the Asian giant tree squirrel are each other's closest relatives, but it is possible, considering that they have probably evolved independently for approximately 30 million years. There is, however, little question that these two groups are ancient lineages, because they both have anatomical peculiarities, notably in skull morphology, that have long been recognized.

The other squirrel subfamilies diverged later but still in the late Oligocene or early Miocene. The subfamily Sciurinae includes all flying squirrels (tribe Pteromyini) and all tree squirrels (tribe Sciurini) of North America, Europe, northern Asia, and South America (except the pygmy squirrel)—a total of 81 species. It has been debated for a long time whether gliding evolved once or twice among squirrels. The molecular data in both investigations clearly support the thesis that gliding evolved only once in the family Sciuridae. This conclusion agrees with the anatomical data, but the close relationship between tree squirrels and flying squirrels is not in accord with some interpretations of the fossil record of flying squirrels. We think that this calls for reinterpretation of the fossil record (which unfortunately is based almost entirely on teeth) because there are some striking anatomical similarities between the Pteromyini and the Sciurini.

A group of tree squirrels in southern Asia forms a fourth subfamily, the Callosciurinae, which includes 64 species. Callosciurinae means "beautiful squirrels," and some species of these, such as the tricolored Prevost's squirrel of Malaysia, rival the dramatic coloration of the Indian giant squirrel. This subfamily was first recognized and described by Reginald Pocock at the British Museum of Natural History in 1923, based on his study of the penis bones of squirrels. His conclusions are completely concordant with the results of the two DNA studies. The fifth subfamily, Xerinae, is the largest group and includes 128 species. Almost all the ground squirrels belong to it: the North American and Eurasian ground squirrels, marmots, chipmunks, and so on (tribe Marmotini), and the African ground squirrels (tribe Xerini). The subfamily Xerinae also includes the 30 species of African tree and bush squirrels (tribe Protoxerini). There has been some speculation that two different squirrels invaded Africa, with one the progenitor of the African ground squirrels and the other the progenitor of the African tree and bush squirrels. The molecular data support this hypothesis.

Squirrels: The Animal Answer Guide

Table 1.1. Classification of the Sciuridae

order **RODENTIA**
family **Sciuridae**
    subfamily **Sciurillinae** (1 species)
        South American pygmy squirrels
    subfamily **Ratufinae** (4 species)
        giant tree squirrels of southern Asia
    subfamily **Sciurinae** (81 species)
      tribe **Sciurini**
        North and South American, northern Asian, and
        European tree squirrels
      tribe **Pteromyini**
        all flying squirrels
    subfamily **Callosciurinae** (64 species)
        southern Asian tree squirrels
    subfamily **Xerinae** (128 species)
      tribe **Marmotini**
        North American and Eurasian ground squirrels
      tribe **Xerini**
        African ground squirrels
      tribe **Protoxerini**
        African bush and tree squirrels

What characterizes the major groups of squirrels?

The giant tree squirrels of southern Asia (subfamily Ratufinae) are twice to three times the size of the largest tree squirrels elsewhere. They frequently average 1,800 grams (4 lbs) and individuals may exceed 2 kg (4.4 lbs). To compare, the largest tree squirrels in North America, the fox squirrels, weigh approximately 1 kg (2.2 lbs), the largest squirrels of South America are somewhat smaller, and the largest tree squirrels of Africa are only 600–700 grams (1.3–1.7 lbs). The Ratufinae have extraordinarily long tails, which are used to counterbalance the mass of the body. They are impressive animals, and visitors to the Smithsonian, who are not familiar with the giant tree squirrels, are always awed when we show them specimens that are more than 76 cm (30 inches) long, from tip of nose to tip of tail. In the wild, the squirrels frequently sit crosswise on a branch with their body hanging down on one side and their tail on the other side, instead of sitting up on their haunches like other tree squirrels. The giant squirrels also have larger thumbs and manipulate their food with them more than do other tree squirrels.

The South American pygmy squirrels (Sciurillinae) are small (33–45 grams; 1.2–1.6 oz), active, normal-looking squirrels, at least on the outside.

The Indian giant squirrel (*Ratufa indica*) shows the feeding posture characteristic of the giant tree squirrels. Photo © Sudhir Shivaram

They are the smallest South American squirrels, about the same size as the least chipmunk (*Tamias minimus*) of North America. Smithsonian researcher Louise Emmons describes them feeding from close to the ground to high in the trees in the rain forest. They are bark gleaners, which forage on tree trunks, prying or gouging small pieces of loose bark, and scraping food off the undersurface. Anatomically, they are very distinct. When he was a researcher at the American Museum of Natural History, the late Joseph Moore considered that these squirrels were some of the most unusual squirrels and listed 12 distinctive features of skull morphology, noting that the South American pygmy squirrels differ from every other group in at least six of these features.

The common tree squirrels of southern Asia are the squirrels in the subfamily Callosciurinae, which were formerly confused with the American and the Holarctic tree squirrels (tribe Sciurini of the subfamily Sciurinae). In 1923, Pocock reported on his study of the baculum (the penis bone) of squirrels. He demonstrated that this bone is very useful for sorting out the taxonomy of squirrels. He found that all Callosciurinae have an extra blade on the baculum seen in no other squirrels. This feature evolved in the lineage of this group, shortly after it diverged from the lineage of the Indian striped squirrels (*Funambulus*), just as suggested by Pocock. Recent molecular studies of squirrel DNA have completely supported Pocock's

Squirrels: The Animal Answer Guide

The tricolored Southeast Asian Prevost's squirrel (*Callosciurus prevostii*). Photo © Jesse Cohen, National Zoo

findings, and the subfamily is now considered to include *Funambulus* and all the other tree squirrels of southern Asia, except of course, the Ratufinae. You do not need to dissect one of the squirrels to distinguish it from an American or Holarctic tree squirrel. In a zoo or elsewhere, just look at the size of the ear. The Callosciurinae usually have much smaller external ears than the North American or Eurasian tree squirrels, and only one, the diminutive pygmy squirrel, *Exilisciurus whiteheadi*, has tufted ears. Many species of the Callosciurinae are very distinctive, like the dramatic tricolored Prevost's squirrel (*Callosciurus prevostii*), the ant-eating, long-nosed shrew-faced squirrel (*Rhinosciurus laticaudatus)*, and the pygmy black-eared tree squirrel (*Nannosciurus melanotis)* with its white mustache mark. Others are less distinctive, like the Indian striped squirrels (*Funambulus*) and the striped squirrels of Southeast Asia (*Tamiops)*, which look like North American or Asian chipmunks (*Tamias*).

The Sciurinae is the subfamily that includes the flying squirrels (tribe Pteromyini), the tree squirrels of the Americas (excluding the South American pygmy squirrel), the tree squirrels of Europe and northern Asia, and one squirrel (*Rheithrosciurus*) of Borneo (tribe Sciurini). All Pteryomyini are principally nocturnal. They range in size from the 15-gram (approximately half an ounce) pygmy flying squirrel (*Petaurillus emiliae*) on Borneo to the 2.5-kg (5.5 lb) woolly flying squirrel (*Eupetaurus cinereus*) of the mountains of north Pakistan and Tibet. They are widely distributed in North America, across northern Eurasia, from Finland to Japan, and are most speciose in southern Asia, in particular, in Borneo.

The Pteromyini are easily recognized by the skin extending from the wrist to the ankle, the patagium, which serves as a wing or parachute when they glide. Larger flying squirrels also have skin extending from the ankle

The northern palm squirrel (*Funambulus pennantii*) from South Asia has adapted well to living in a human-altered environment. Photo © Dr. S. N. Naik

to the tail, the uropatagium, that acts like a rear flap on a plane's wing. Small flying squirrels don't have much, if any, of this rear flap.

There are distinctive marks on the wrist and ankle bones of flying squirrels associated with the attachment of the patagium muscles. In the wrist there is a special arrangement of three bones (the pisiform, triquetrum, and the scapholunate), which provides increased stability at the wing tip, at the cost of reduced mobility of the wrist. Just above the ankle, on the long bone (the tibia), many flying squirrels have a small extra bump for the attachment of a special wing muscle. On other flying squirrels, this muscle attaches on the foot, but there is no bump there. Smaller flying squirrels (less than 1 kg) usually have tails with long lateral hairs (a pattern called distichous), but bigger flying squirrels have relatively longer tails that are round (a pattern called terete)—with lateral hairs the same length as those above and below.

The tree squirrels (tribe Sciurini) are more difficult to characterize. Their geographic range includes South America, North America, Europe, northern Asia, and a single species in Borneo. They are typical tree squirrels in having usually long, distichously haired tails, but in contrast with most of the Callosciurinae, almost all species have larger ears, which are frequently tufted. The Sciurini also lack the extra blade on the baculum that all Callosciurinae possess. The bacula of the South American pygmy squirrel (*Sciurillus*) and squirrels in the genus *Sciurus*, however, are very similar, suggesting that the bacular morphology of the Sciurini is similar to the ancestral form. In 1959, Joseph Moore listed four somewhat distinc-

Squirrels: The Animal Answer Guide

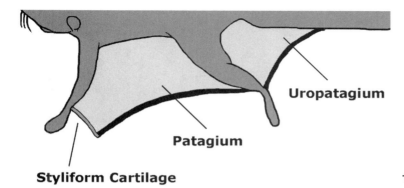

Uropatagium

Patagium

Styliform Cartilage

The basic parts of a flying squirrel

tive features of the skull in the Sciurini, but cranial anatomy is not very diagnostic. There is an interesting and distinctive feature of the ankle joint that characterizes both the flying squirrels and the tree squirrels of the subfamily Sciurinae. There is normally a prominent sulcus or groove on the underside of the first ankle bone, the astragalus. It is quite prominent in the Ratufinae and the Callosciurinae, and only slightly less prominent among the African tree squirrels. However, the sulcus is nearly obliterated in the Sciurini tree squirrels and it is even less evident in the flying squirrels. We think it has to do with ankle rotation and the way these squirrels turn their feet around when descending a tree head first, because obliteration of the sulcus is associated with an increased joint surface between the astragalus and the calcaneus, where ankle rotation occurs. Chipmunks and some other North American ground squirrels that climb have the sulcus obliterated also, however, complicating the interpretation and diagnostic utility of this feature.

The subfamily Xerinae includes such a heterogeneous group of squirrels that it is the most difficult of all subfamilies to characterize. Included are the African ground squirrels (tribe Xerini), all the ground squirrels of North America and northern Eurasia—from the chipmunks to the marmots and prairie dogs (tribe Marmotini)—and all the African tree squirrels, including the sun squirrels, the rope squirrels, and the bush squirrels (tribe Protoxerini). Only molecular biology unites these three tribes into a single subfamily, but on the basis of two separate studies utilizing five different genes. Joseph Moore did note five cranial characters shared by squirrels of the tribes Xerini and Marmotini, but he considered these similarities to be due to convergent evolution not common ancestry.

The tribe Xerini includes four species of African ground squirrels and the long-clawed ground squirrel (*Spermophilopsis leptodactylus*) from Turkmenistan, Uzbekistan, and neighboring countries. *Spermophilopsis* and *Xerus rutilus* are unstriped, whereas the other species have prominent white stripes along their sides. *Spermophilopsis* is the only one with a short tail. All four species share seven distinctive cranial features, four of which distin-

The unstriped ground squirrel (*Xerus rutilus*) of East Africa. Photo © U.S. Fish and Wildlife Service

guish them from the Marmotini. One of these is a bony palette that extends well behind the cheek teeth, a feature that distinguishes them from most other squirrels.

The tribe Marmotini is difficult to characterize, because it includes animals as different as chipmunks (*Tamias*) and marmots (*Marmota*). However, it has long been recognized as a natural group on the basis of several characteristics. All species, for example, have cheek pouches, which are not found in any other squirrels. Another anatomical feature is the layer of the deltoid muscle that extends from the clavicle to the forearm in the marmots, ground squirrels, and prairie dogs. This is unique among squirrels. The species of Marmotini come in a bewildering variety of colors and patterns. Chipmunks are striped, as are many ground squirrels. Some ground squirrels are spotted, or even striped and spotted (the 13-lined ground squirrel, *Spermophilus tridecemlineatus*). Marmots and prairie dogs are unstriped and relatively uniformly colored. Size differences are also extreme. The least chipmunk may average 40 grams (1.4 oz), but the largest marmots may weigh 8 kg (almost 18 lbs). Tail length is also variable: The tail is 45% of the total length of the Baja California rock squirrel (*Spermophilus atricapillus*), 20–25% of the total length of the woodchuck (*Marmota monax*), and less than 20% of the total length of the white-tailed prairie dog (*Cynomys leucurus*).

The tribe Protoxerini includes the African sun squirrels (genus *Heliosciurus*), rope squirrels (*Funisciurus*), giant squirrels (*Protoxerus* and *Epixerus*), pygmy squirrel (*Myosciurus*), and bush squirrels (*Paraxerus*). The squirrels of the Protoxerini are also diverse and are combined into one tribe by DNA

Squirrels: The Animal Answer Guide

Yellow-bellied marmots (*Marmota flaviventris*) range from southern British Columbia and Alberta south to northern New Mexico. Photo © National Park Service

Prairie dogs, like this black-tailed prairie dog, inhabit the great plains of southern Canada, mid-America, and northern Mexico. Photo © U.S. Fish and Wildlife Service

data but not by similar appearance or morphology. They range in size from the 16-gram (approximately half an ounce) African pygmy squirrel (*Myosciurus pumilus*) to the 700-gram (1.5 lbs) giant forest squirrel (*Protoxerus stangeri*), some are striped (species of *Funisciurus* and *Paraxerus*), most are not, and neither the study of bacula by Pocock nor the study of skulls by Moore suggested a close relationship among them all. It is likely that the genetic studies correctly identify their common ancestry, and the morphological studies document the evolutionary divergence since their ancestor invaded Africa, probably 20 million years ago.

Introducing Squirrels

Golden-mantled ground squirrel (*Spermophilus lateralis*) with stripes and white eye ring, characteristic of some ground squirrels. Photo © National Park Service

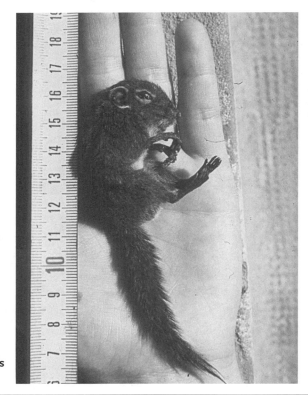

Weighing only 16 grams (or slightly more than half an ounce), the African pygmy squirrel (*Myosciurus pumilio*) is one of the smallest squirrels in the world. Photo © Alain Devez

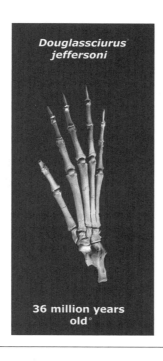

Douglasciurus jeffersoni

36 million years old°

Tamiasciurus hudsonicus

Modern

A comparison showing the relative similarities between the rear foot of the 36-million-year-old *Douglasciurus jeffersoni* and the rear foot of a modern North American red squirrel (*Tamiasciurus hudsonicus*). Notice how the fourth digit of both squirrels is the longest, which is typical of tree squirrels.

When did squirrels evolve?

Squirrels evolved from more primitive rodents approximately 36 million years ago, at the end of the Eocene or the beginning of the Oligocene epoch. At this time, the earth was much warmer than it is now, and consequently it was more heavily forested. This probably provided an arboreal niche, for which the ancestral squirrels became adapted.

What is the oldest fossil squirrel?

There is a difference of opinion about which is the oldest fossil squirrel, resulting more from differences in definition than from disagreement about the fossils. Jefferson's squirrel (*Douglasciurus jeffersoni*) dates from the late Eocene of Wyoming, 36 million years ago. It is an extraordinarily complete skeleton, found inadvertently during a 1975 field season by Jennifer Emry, the wife of Smithsonian paleontologist Bob Emry. Jefferson's squirrel has a skeleton surprisingly like that of a modern tree squirrel. In fact, Thorington mistook it for a modern squirrel when he first saw it. However, the skull of Jefferson's squirrel exhibits an ancestral anatomy of the jaw muscles, not the anatomical arrangement found in all later fossil squirrels and all modern squirrels. Therefore, some paleontologists exclude it from the squirrel family, although they may consider it to be ancestral or closely related to the ancestor of the family. Others, like us, consider it to be a squirrel despite the primitive jaw musculature. There is no question, however, that it is the oldest fossil.

The next oldest fossil, described by French paleontologist Monique Vianey-Liaud in 1974, is the 31-million-year-old *Paleosciurus goti* from the early Oligocene of southern France. It is also a surprisingly complete skeleton and is the oldest fossil to exhibit the typical jaw musculature of squirrels, hence the claim that it is the oldest fossil squirrel. Unexpectedly, on the basis of the relative lengths of the arm bones, some have claimed that it was a terrestrial squirrel. We are uncertain about this claim, however, because the bones look very much like the bones of a tree squirrel, based on the published illustrations. Also, the illustration of the radius shows that this forearm bone is slightly damaged at one end, causing it to be shorter than expected of a tree squirrel and more like a ground squirrel, as reported.

Form and Function

What are the largest and smallest squirrels?

The largest squirrels are the marmots (*Marmota*), which accordingly can be called "the giant ground squirrels." In North America, the best known are the yellow-bellied marmot of the Rocky Mountains and the woodchuck of the eastern United States and Canada, which is the proverbial "ground-hog" of February 2. The largest of all is the gray marmot found in the mountains of Khazakstan. All marmots put on weight before they enter hibernation—some of them double their weight—so the animals are heaviest at the end of the summer. At this time, the largest gray marmots may weigh more than 8 kg (18 lbs), truly a giant squirrel! Among tree squirrels, the giant tree squirrels of Southeast Asia (*Ratufa*) are not nearly as big as the marmots, but they are still quite large, ranging from approximately 2 to 3 kg (4 to 6 lbs). With their beautiful long tails and striking coloring, these squirrels are impressive as they bound above you through the trees.

In contrast, the smallest squirrels are the pygmy tree squirrels, such as the ones in western Africa (*Myosciurus pumilio*) and the ones in Southeast Asia (*Exilisciurus* spp.), which are smaller than some mice. The smallest adults of both genera average approximately 14 or 15 grams (approximately half an ounce). They are so tiny you could mail two of them first class from New York to Los Angeles for a 39¢ stamp.

How fast does a squirrel's heart beat?

Mammalian heart rate is inversely related to body size, thus smaller squirrels have faster heart rates than larger ones. For example, the 13-lined

African Pygmy Squirrel

The skull of the largest and smallest tree squirrels in Africa. (Top) The African pygmy squirrel (*Myosciurus pumilio*), weighing approximately 15 grams. (Bottom) The forest giant squirrel (*Protoxerus stangeri*), weighing approximately 600 grams.

Forest Giant Squirrel

ground squirrel, weighing approximately 140 grams (5 oz), has a heart rate of about 280 beats per minute. The heart of a 15-gram (half an ounce) pygmy squirrel probably beats about 500 times each minute, whereas the heart of an 8-kg (17.6 lbs) gray marmot (nonhibernating) is expected to beat approximately 145 times a minute, although neither has been measured. (For comparison, a human heart beats about 75 times each minute.) The heart rate of a hibernating squirrel can be amazingly slow, dropping to only 3 to 15 beats per minute. An individual squirrel's heart rate will vary, and a chipmunk fleeing a hawk, unsurprisingly, will have a much higher heart rate than a one sunning itself on a log.

Can squirrels see color?

Yes. Squirrels have dichromatic color vision. They can distinguish color much like a human who has red-green color blindness—which means they can differentiate red or green from other colors, but cannot distinguish red and green from each other.

Several other aspects of squirrel vision are worth noting. Many squirrels have yellow-tinted eye lenses. Ground squirrels have dark-yellow lenses and tree squirrels have paler-yellow lenses. These yellow lenses, much like

Squirrels: The Animal Answer Guide

sunglasses, reduce glare from bright light and increase the contrast between colors, giving the squirrel sharper vision. Flying squirrels, however, have clear lenses. Because they are nocturnal and seldom encounter bright light, they have no need for a tinted lens.

Squirrels also have exceptional focusing ability. Human eyes have a fovea centralis or a small area of the retina where cones are most densely packed and vision is most acute. This is the part of the retina you use when you read. Squirrels, on the other hand, have sharp vision across the entire retina, which allows a motionless squirrel to see clearly what is next to it and above it at the same time without moving its head. Thus, a squirrel could read the small print of a newspaper with its peripheral vision.

As in most mammals, the squirrel retina contains both rods and cones. Rods are specially adapted cells that enable vision in low light, and cones are specially designed cells for daylight vision, color vision, and the discrimination of detail. That the retina of diurnal squirrels contains both rods and cones makes sense, because although they are primarily active during the day, they also need to see at dusk and dawn and in shaded areas. Ground squirrels, such as the prairie dog, are superbly adapted to bright light and have many more cones than rods. In fact it was once thought that they had no rods at all. The retinas of the nocturnal flying squirrels, on the other hand, have mostly rods and few cones, which gives them excellent night vision.

Good vision is very important to squirrels. It helps tree squirrels safely navigate through a complex three-dimensional environment and helps social ground squirrels identify and interact with each other. Most importantly, good vision is crucial for spotting and avoiding potential predators.

Do all squirrels have cheek pouches?

No, cheek pouches of various sizes are found only in ground squirrels—specifically in the chipmunks, antelope ground squirrels (*Ammospermophilus* spp.), rock squirrels, and other ground squirrels (*Spermophilus* spp.), marmots, and prairie dogs. No tree squirrels or flying squirrels have cheek pouches and neither, surprisingly, do any of the African ground squirrels.

There is a muscle in the cheek pouch that helps with emptying it, and this muscle attaches to the squirrel's skull just behind the upper incisors. This muscle leaves a prominent scar on the bone where it attaches. When paleontologists find a fossil squirrel skull, they can infer that the squirrel had a cheek pouch if this scar is present. Accordingly, they have been able to document that cheek pouches evolved very early in the history of ground squirrels. From this evidence, we assume that early ground squirrels collected small seeds, carried them in their cheek pouches, and probably hoarded food in their burrow, the way chipmunks do today.

A foraging eastern chipmunk (*Tamias striatus*) fills its cheek pouches with sunflower seeds, which it will bring back to its burrow for storage. Photo © Phil Myers

Marmots, which evolved more recently and do not store food, have smaller cheek pouches than other squirrels in their lineage. We presume that the habit of feeding on the softer, vegetative parts of plants, the way marmots do, evolved later and led to the reduced usefulness of a cheek pouch and its reduction in size.

Can squirrels swim?

Yes, there are many records of tree squirrels swimming. During their migrations, eastern gray squirrels may swim across relatively large rivers, as commonly reported in past centuries. Bill Hamilton, at Cornell University, reported an abundance of squirrels in Connecticut and New York in the fall of 1933, with more than a thousand squirrels estimated to have attempted to swim across the Connecticut River near Hartford. During September 1968, population densities of eastern gray squirrels were very high throughout the eastern United States, and Vagn Flyger, at the University of Maryland, reported that squirrels were seen crossing rivers and reservoirs in a number of areas, including reservoirs in Tennessee and the Connecticut River near Hartford. In the autumn of 1990, there were dense populations of eastern gray squirrels in the vicinity of Washington, D.C., and they were observed swimming across the Potomac River. Similar observations have been reported for the Eurasian red squirrel. Robert Hatt cited several

Squirrels: The Animal Answer Guide

A woodchuck (*Marmota monax*) forages along the water of the Tred Avon River off of the Chesapeake Bay. Although reports of ground squirrels swimming are rare, it is probable that those living near water do wade into the water to get vegetation. Photo © Peter Wainwright Thorington

accounts of swimming in the North American red squirrel. Most recently, in 2005, Jonathan Pauli of University of Wisconsin-Stevens Point was kayaking in Lake Superior and noticed a North American red squirrel swimming. He followed the squirrel as it swam 30 minutes, about 1.5 km (approximately 1 mile) from the mainland to a nearby island through choppy water. This is the first report of a squirrel swimming in the Great Lakes, as well as the longest reported swim of a North American red squirrel.

Reports of ground squirrels swimming are more scanty. In 1883, renowned American zoologist C. Hart Merriam observed a juvenile woodchuck swimming across a lake in the Adirondacks but considered this an unusual observation. Golden-mantled ground squirrels have been seen swimming across small streams, and under experimental conditions, Richardson's ground squirrels swim capably. All ground squirrels probably can swim but few choose to do so. In contrast, flying squirrels are not good swimmers, and there are some reports of them drowning when they landed in water.

How far can squirrels jump?

This is one of those simple questions that is very difficult to answer. Nature writer John Burroughs described a squirrel that was released at

An Indian giant squirrel (*Ratufa indica*) makes a leap in Bandipur National Park, Bandipur, Karnataka, India.
Photo © Sudhir Shivaram

the edge of a canyon and leaped off into space. He peered over the edge to see that the squirrel had safely caught itself on a small ledge, far below. That squirrel probably accomplished a world record leap, but it is not one that we would count because it was really more of a fall. Therefore, we must define the question better, perhaps by adding "without losing height"— meaning the squirrel must land on an object at the same height as its take-off point. Reports in the literature about distances jumped are very rarely that specific, so our answer is that we really do not know how far squirrels can jump.

Thorington approached this problem experimentally with the eastern gray squirrel by placing a feeding platform full of seed beside a stump. Gradually, the platform was moved farther and farther away from the stump. Because it was a hanging platform and too high for them to reach by leaping from the ground, the squirrels had to climb the stump and then leap to the feeding platform. As the distance between stump and platform increased, the jump became an elimination contest. First the younger squirrels and then others refused to jump and instead would forage under the platform. Of those squirrels that jumped, he never saw one miss—they seemed to judge their limitations and refuse to try. Eventually, only a single squirrel was making the leap, at a distance of more than 2.5 m (8.2 feet), approximately 10 times the length of its head and body!

His intention was to conduct a comparative study of leaping in different types of squirrels. When he set up a similar experiment at a campground in California to test the jumping abilities of chipmunks, he encountered a problem. A deer would always find the feeding tray before the squirrels did and lick it clean. So his experimental work is incomplete. He became involved in another project and we still do not know how far different species of squirrels can jump.

Squirrels: The Animal Answer Guide

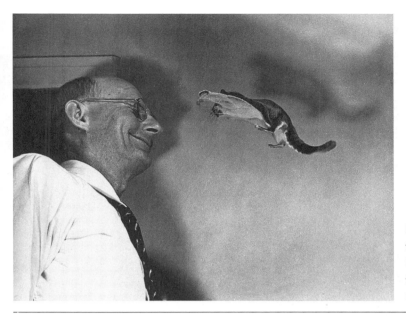

Do squirrels fly?

The term *flying squirrel* can be misleading. Flying squirrels do not fly the way bats and many birds do. Instead, they glide. With the aid of a special membrane, they can glide from a higher perch to a lower one. The physics behind the gliding of the flying squirrels is fascinating. By starting at a high point and moving toward a low point, a gliding squirrel converts the energy created by the vertical drop into forward movement and can glide three to four times as far as it drops. In other words, a flying squirrel can glide half the length of a football field from a 15 m (50 feet) tall tree. Under some conditions, in particular, if there is an updraft, they can go farther. Gliding, however, is not the only means of locomotion that the flying squirrels employ. Flying squirrels can race through trees and bound across ground much like nonflying tree squirrels.

Flying squirrels are not the only mammals that glide. At least six groups of mammals, from marsupial possums to colugos, have evolved gliding flight, but these other gliding species do not enjoy the wide geographic distribution of the flying squirrels.

What are the largest and smallest flying squirrels?

The largest flying squirrel, and in fact the largest mammalian glider, is the woolly flying squirrel (*Eupetaurus cinereus*) of northern Pakistan, Afghanistan, and northwestern India. This is a poorly known species, measuring from 900 to 1,200 mm (35 to 39 inches) from head to tail; adult animals of up to 2.5 kg (5.5 lbs) have been weighed. Some of the so-called giant flying squirrels (*Petaurista* spp.) are almost as large.

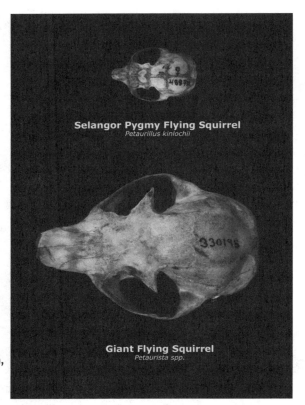

The skulls of some of the largest and smallest flying squirrels. (Top) The Selangor pygmy flying squirrel (*Petaurillus kinlochii*), weighing approximately 19 grams. (Bottom) A giant flying squirrel (*Petaurista* spp.), weighing approximately 2,000 grams.

The smallest flying squirrel is undoubtedly the pygmy flying squirrel, *Petaurillus emiliae*, of Borneo, which probably rivals the pygmy tree squirrels in diminutiveness, but it is known only from two specimens, which were not weighed. Thorington examined these specimens in the British Museum and found them considerably smaller than the closely related species, *Petaurillus kinlochii*, which weighs 19 to 22 grams (0.7 to 0.8 oz).

How do you make a flying squirrel?

We excluded a more detailed discussion of flying squirrel anatomy from the "What are squirrels?" question, because flying squirrels are special, and their anatomy includes unique features that can best be explained in the context of the question "How do you make a flying squirrel?"

WINGS. The obvious first answer to our question is that a flying squirrel must have wings, but flying squirrels do not have wings like birds. Instead, they glide with a parachute-like membrane, called a patagium, which extends from the forelimb to the hind limb. Large flying squirrels (more than 1 kg or 2 lbs) have an additional membrane, called a uropatagium, between the hind limb and the tail. The patagium and uropatagium are made up of

32

Squirrels: The Animal Answer Guide

two layers of skin with fur on the outside and a thin layer of muscles and nerves in between. When the squirrel is not gliding, it uses these muscles to gather the membrane close to the body, out of the way, so that it can run and climb easily. There are also thin cordlike muscles bordering the patagium. One of these extends from the wrist to the ankle, and in some flying squirrels there is a small bump on one of the leg bones, the tibia, where it inserts. This is of potential interest to paleontologists for identifying fossil bones. Of interest to anatomists is how one of the thigh muscles lines the edge of the uropatagium in large flying squirrels. By comparing this muscle in different flying squirrels, we can see how it has evolved from a muscle of the thigh to a completely separate muscle, extending from the ankle to the tail in the largest flying squirrels.

Long limbs. Flying squirrels have the longest limbs relative to body size of all the squirrels and this appears to be a necessary part of the flying squirrel adaptation. If you stretch a patagium between the limbs of a modern tree squirrel, you get a rectangular shape. If you elongate the squirrel's arm and leg bones the patagium becomes square shaped. This change in shape is important, because it affects the gliding ability of the squirrel. A rectangular-shaped patagium does not provide as much surface area, resulting in less lift and more drag, whereas a square-shaped patagium enables the squirrel to glide further horizontally relative to the distance it drops vertically. This shape also allows the squirrel to land at a relatively slow speed and with a high angle of attack (with its head up and feet forward), instead of a low angle of attack (head first).

Steering and stabilization. Being able to glide longer distances is only one part of the adaptation. A flying squirrel must also be able to steer and to land safely. It appears that small flying squirrels use their tails to steer but that large flying squirrels probably steer by adjusting their arm positions. Although the aerodynamics of steering and landing are not well known for flying squirrels, stability clearly plays an important role.

Greater wing tip stability in flying squirrels is accomplished by reducing movement between the forearm bones, the radius and the ulna, and the three wrist bones, the scapholunate, triquetral, and the pisiform. This stabilization should come with a cost, however, because it should compromise manual dexterity. To understand this, sit in front of a table and place your hand palm down on the table. Now, turn your hand palm up. This movement is called supination. Now, turn the palm down again. This movement is called pronation. If a surgeon were to pin your radius and ulna together near the wrist, you would not be able to pronate or supinate your hand, which would greatly reduce your manual dexterity.

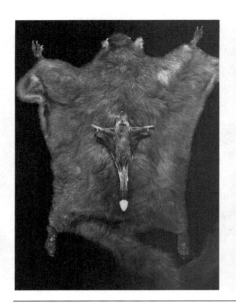

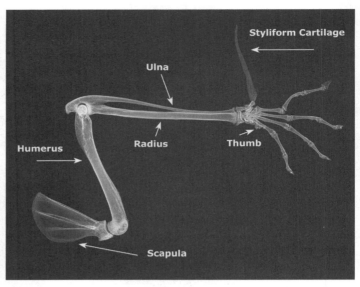

(Left) A museum specimen of the Selangor pygmy flying squirrel lies on top of a giant flying squirrel, illustrating the range in body size among flying squirrels. (Right) An x-ray of the complete arm of a southern flying squirrel (*Glaucomys volans*). Note the styliform cartilage, where the gliding membrane attaches. Also see the minuscule thumb and how the radius and ulna are bound together near the wrist.

If we look at the anatomy of the southern flying squirrel (*Glaucomys volans*) we see that (1) the radius and ulna are tightly bound together at the wrist and (2) the pronator and supinator muscles of the forearm are large. Finally, if we observe living flying squirrels it is apparent that they have lots of manual dexterity and can pronate and supinate their hands readily. We have determined that pronation and supination are occurring at the elbow instead of the wrist in these flying squirrels, with the radius and ulna rotating together as a whole. This is an extraordinary way to increase wrist stability without compromising manual dexterity.

The tails of flying squirrels also are important for steering and stabilization. Small flying squirrels (smaller than 1 kg or 2 lbs) have broad tails with long hairs on each side (distichous), much like the tails of many tree squirrels. Larger flying squirrels (larger than 1 kg or 2 lbs) have long round tails, with all hairs of equal length. In large flying squirrels, the base of the tail, for a short distance, is connected by the uropatagium to the hind legs. We think that the broad tail of a small flying squirrel is used much like the rudder of a boat, whereas the uropatagium in large flying squirrels is probably used more like the flaps on the back edge of an airplane wing.

TURBULENCE. A gliding squirrel needs to be able to deal with air turbulence. If you have ever watched a kite or a glider come crashing to the ground, you have probably observed the devastating effects of air turbu-

Squirrels: The Animal Answer Guide

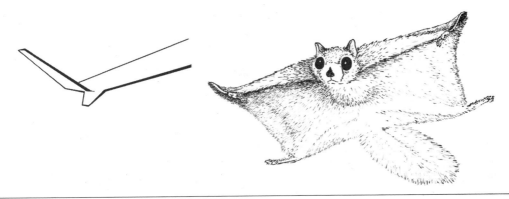

The flying squirrel flies with the tips of the patagium turned up, much like the wingtips of some airplanes. This affords the squirrel greater stability and steering ability. Photo © Karolyn Darrow

lence. An airfoil, however, like a kite or gliding squirrel, can be self-correcting in the face of air turbulence, if its wing tips point slightly upward. This is because of the dihedral effect—when a dihedral-shaped airfoil is tipped too far to the right, the right side provides more lift and tips it back to the left; if it is tipped too far to the left, the left side provides more lift and tips it back again to the right.

Flying squirrels glide with the tips of their patagium turned up and so benefit from the dihedral effect. How they go about elevating the tip of their patagium is pretty ingenious. The tip of the patagium is supported by a flexible rod of cartilage, which is attached at the wrist to the pisiform bone. This cartilage holds the tip of the patagium up when the wrist is cocked toward the thumb side (radially abducted) and then flexed backward (dorsiflexed). Anatomists had been puzzled by the fact that flying squirrels have a very small thumb, like tree squirrels, but a prominent muscle for abducting the thumb. Examining the hand anatomy carefully, we found a ligament that extends across the palm from the thumb to the pisiform bone, so that when the abductor muscle pulls on the thumb it also pulls on the base of the tip of the patagium, causing it to be extended and raised. This seems like a bizarre arrangement, whereby a muscle of the thumb controls the tip of the patagium on the other side of the hand. We can see how it evolved from the anatomy of a tree squirrel hand in which no other muscle is positioned to extend and raise the tip of the patagium.

How far can flying squirrels glide?

It seems that most flying squirrels can glide a distance at least three times as far as the distance they drop in height. Thus, the longest glides are initiated at the greatest heights—the tallest trees, the tops of hills, cliff tops, etc. Under normal circumstances in the wild, accurate measurements of the

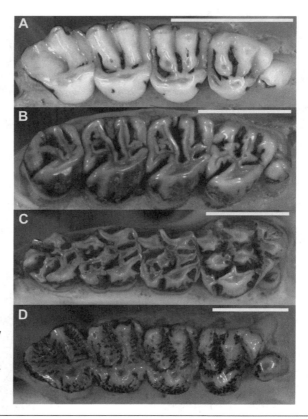

Upper tooth rows of four species of flying squirrels showing the diversity of tooth patterns. (A) *Eoglaucomys fimbriatus.* (B) *Petaurista petaurista.* (C) *Trogopterus xanthipes.* (D) *Aeromys tephromelas.*

distances covered are difficult to obtain—the observer is never close to both the origin and termination of the glide. We do have some facts about the distances of observed glides.

Good data are available on the gliding of the small North American flying squirrels, both the northern species, *Glaucomys sabrinus*, and the southern species, *Glaucomys volans*. John Scheibe and his students at Southeast Missouri State University have studied the southern flying squirrel in Missouri; they report the lengths of several glides, with a maximum of 45 meters (148 feet). Karl Vernes and his students at Mount Allison University studied northern flying squirrels in New Brunswick and recorded the distances the squirrels glided in the wild, when released from their live traps. The males averaged 19 meters (62 feet) and the females, 14 meters (46 feet), but the maximum distance was also 45 meters (148 feet).

The large flying squirrels in southern Asia can glide phenomenal distances. The Japanese flying squirrel (*Petaurista leucogenys*) has been recorded to glide 115 meters (377 feet). The squirrels do not normally seem to choose to glide long distances. When possible, they prefer to make a series of shorter glides rather than one long glide. The average glides recorded by Japanese biologists Motokazu Ando and Satoshi Shiraishi at different study sites ranged between 17 meters (56 feet) and 33 meters (108 feet).

Squirrels: The Animal Answer Guide

Peter Zahler, while at the University of Massachusetts, studied the woolly flying squirrel (*Eupetaurus cinereus*) in northern Pakistan, and on one occasion he estimated a glide to be 150 meters (492 feet), with a 50-meter (164 feet) drop. Again, the average glides for this species were much shorter.

We have the uncomfortable feeling that we are underestimating the capabilities of the large flying squirrels, because there are too many stories about glides too long to be measured. Illar Muul, formerly of the U.S. Army Medical Research Unit in Malaysia, has described flying squirrels (*Petaurista petaurista* and others) living in caves in limestone cliffs above a village in Malaysia and gliding from these caves in the evenings down to their feeding trees near the village—distances considerably longer than those listed above. Similarly, Smithsonian researcher Brian Stafford saw some very long glides of the Japanese flying squirrel that he was unable to measure.

How can you determine whether a fossil squirrel was a tree squirrel, a flying squirrel, or a ground squirrel?

The surest way to determine if a fossil squirrel was a tree squirrel, a flying squirrel, or a ground squirrel is to study the limb bones. Unfortunately, most fossil squirrels are represented only by teeth and jaws, without associated limb bones. When limb bones are present, flying squirrels can be recognized by their long and slender bones, tree squirrels by their shorter bones with more prominent muscle attachments, and ground squirrels by their more robust bones with extremely prominent muscle scars.

In flying squirrels the radius and ulna are tightly bound together at the wrist, and in some flying squirrels there is a small bump on one of the leg bones, the tibia, where the muscles lining the patagium insert. These characters may be diagnostic for paleontologists.

Some small bones can also be very distinctive, such as the wrist bones of flying squirrels or the short robust finger bones of ground squirrels. Anatomical features associated with the stabilization of a flying squirrel's patagium also should be evident in the fossil record, if the relevant bones are present, and may enable paleontologists to identify early flying squirrels.

Teeth and jaws, although not as diagnostic as limb bones, can also be indicative of the way of life of a fossil squirrel. Paleontologists frequently use teeth to help create a more complete picture of how fossil squirrels were related to one another.

Chapter 3

Coat Color and Squirrel Genetics

What species are the black squirrels I see?

If you are in the eastern United States, it is probably the melanistic form of the eastern gray squirrel (*Sciurus carolinensis*). The black color morph of the eastern gray squirrel occurs most commonly in northern latitudes, but it has been introduced into more southern areas across the United States and has established itself successfully on college campuses and in urban and suburban communities.

Individual variation in coat color is well known in squirrels. In urban and suburban areas, there are frequently several color forms of the eastern gray squirrel. In the vicinity of Washington, D.C., we have black squirrels, albino squirrels, the normal gray squirrels, reddish squirrels, and a silvery squirrel, which can look albino at a distance, but close up can be seen to have dark eyes and silvery, not white, hair. All these color forms belong to the same species and subspecies, *Sciurus carolinensis pennsylvanicus*, and can be seen to breed with one another.

The melanistic squirrels found in the Washington, D.C., area have an interesting history. In 1902 and 1906, a total of 18 black squirrels, because they were considered "unusual," were brought to the National Zoo from Canada, where all-black squirrels are common. In 1923, it was noted in the *Mammals of the District of Columbia* that the black coloration had spread and that black squirrels could be found well outside the zoo grounds in Cleveland Park. In the Smithsonian collection there are several specimens documenting this, all collected by the famous anthropologist, Aleš Hrdlička, who was probably picking up roadkill squirrels on his way to work. When Thorington came to Washington in 1969, he saw black squirrels regularly,

Two eastern gray squirrels eat at a stump in Thorington's backyard: on the left, a normal colored one, and on the right, a melanistic one. Photo © Caroline Thorington

but always inside the Washington Beltway, and he never heard any reports of them in Virginia, across the Potomac River. In the 1980s, he began to see black squirrels outside the Beltway in Maryland, and he began to receive phone calls from long-term Virginia residents, inquiring about the black squirrels that they were seeing for the first time. Now, the black squirrels are common in both places, but we do not know how far outside the Beltway or how far south in Virginia black squirrels can be seen. We have received reports of black squirrels 30 miles to the north around Fort Meade, Maryland, 30 miles to the east in Annapolis, Maryland, and 25 miles to the south in Prince William County, Virginia. They are not common in these areas, however, and it is unclear if their presence there results from the natural spread of the gene for melanism from the original introduction at the National Zoo. Melanism can occur naturally in wild populations, and additional introductions are also possible explanations.

The obvious question is why melanism is so successful in urban and suburban environments, such as Washington, D.C.? One hypothesis is that they survive better because motorists are more likely to try to avoid them. Katherine and Richard Thorington tested this hypothesis by counting the numbers of black and gray squirrels they saw living in different neighborhoods in 1990, and the number that they saw dead on the roads. The frequency with which they found dead black squirrels was the same as

Coat Color and Squirrel Genetics

their frequency in the population. So the Thoringtons concluded that black squirrels were just as likely to be run over by motorists as the gray squirrels. Another hypothesis is that natural predators are less common in the cities and suburbs than in the countryside and that this allows the black form to survive better there. This is a reasonable idea, because black squirrels are much more conspicuous than the gray squirrels in many situations, but it needs to be tested. Nevertheless, there still needs to be some advantage to being black, even if predators are not selecting against them. This leads us to the question of why melanistic squirrels are most common in the northern latitudes. One hypothesis is that their black coloring allows the squirrels to absorb more heat when they sun themselves in the winter and that this promotes survival. This hypothesis has been tested, but unfortunately the results are ambiguous. Thermoregulation may be the cause, but the genes of melanism affect much more than just coat color, so there may be an advantage to melanism that we have not discovered.

What species is the red squirrel?

The name "red squirrel" is given to several different species. In North America, it is a species of small tree squirrel (*Tamiasciurus hudsonicus*), also called the pine squirrel or the chickaree. Some subspecies are more gray on the back than red, but others are very deserving of the name. In Europe, the red squirrel is the common Eurasian tree squirrel (*Sciurus vulgaris*). It also has subspecies that are not at all red, particularly in Russia, but in the British Isles it is definitely a red squirrel. To avoid confusion, we refer to *Tamiasciurus hudsonicus* as the North American red squirrel and *Sciurus vulgaris* as the Eurasian red squirrel. In other parts of the world, there are red squirrels that do not belong to either of these species. In Colombia, there are subspecies of the red-tailed squirrel (such as *Sciurus granatensis splendidus*), which are dramatically colored tree squirrels, vivid red on the back, with pure white abdomens. Another red squirrel took Thorington by surprise when he was reviewing the Smithsonian's collection of squirrels preserved in fluid. But we must digress for a moment. At the Smithsonian, we call this collection our "Collection of Alcoholics," because the specimens are stored in alcohol. In London, at the Natural History Museum, their collection of specimens in fluid is called their "Spirit Collection." We suspect that both names were adopted without much thought of their alternative interpretations, but perhaps we underestimate the humor of our predecessors. In the Smithsonian inventory of its collection of alcoholics, there was a specimen listed simply as red squirrel. Because we have numerous specimens of the North American red squirrel in the collection, Thorington gave it no thought until he saw the animal and was delightfully surprised to discover that it was Swynnerton's bush squirrel, a red tree squirrel (*Paraxerus vexillarius*) of southeastern Africa. Finally, there are other dramatic red tree squirrels in Southeast Asia, such as *Callosciurus finlaysonii ferrugineus* of Myanmar (formerly Burma) and three other subspecies of *Callosciurus finlaysonii* in Thailand, Cambodia, and Laos.

What causes the different coat colors of squirrels?

Hair is produced in the skin by cells forming a hair bulb in a follicle. During the growth of the hair, pigment cells form a substance called melanin that is deposited in the inner layers of the hair, the cortex and medulla, and give it its color. The melanin may be black or brown (eumelanin) or a variety of paler colors (phaeomelanin), ranging from reds to yellows. The different coat colors are caused by different patterns of eumelanin and phaeomelanin in the hairs.

How are hair colors determined genetically?

The deposition of melanin in the hair is controlled by several genes. This genetic control has not been studied in squirrels, but it is probably very similar to the control in other mammals, which has been particularly well studied in the laboratory mouse. Some of these genes control the movement of the melanin-producing cells into the follicle, others control the kind of melanin that is deposited, the location where it is deposited, whether it is clumped or dispersed, and the shape of the melanin cells in the hair. All of these features influence the apparent color of the individual hairs, and together they determine the coat color of the mammal.

One of the coat color genes is agouti (A). The normal agouti gene has many variants, all called alleles. The normal agouti allele (A) causes the hairs to be banded. As the hair grows, the hair follicle turns on and off the production of different melanins, producing a banding pattern in the hair. The melanocytes switch abruptly from producing eumelanin to phaeomelanin, based on a biochemical change in the follicle. This gives the coat a salt and pepper appearance, which can be an excellent camouflage. It is probably the most common of the agouti alleles in squirrels.

In some species and in some geographic areas, the nonagouti allele (a) is common. This causes the hairs and coat to be all black because eumelanin production is maintained at a constant rate throughout the growth of the hair. Melanistic squirrels are well known in the eastern gray squirrel, the fox squirrel, the Eurasian red squirrel, the North American red squirrel, the eastern chipmunk, several species of marmots, including the yellow-bellied and Vancouver marmots, the 13-lined ground squirrel, and a diverse array

An albino black-tailed prairie dog looks out of its burrow. Because it lacks melanin, it does not have the black tail tip characteristic of the species. Photo © Shirley Curtis, www .scarysquirrel.org

of subspecies of tree squirrels in Southeast Asia, including *Callosciurus prevostii pluto* in Borneo and *Callosciurus finlaysonii nox* in Thailand. It is probable that not all of these are due to the nonagouti allele because other genes can also produce all-black coats. In the eastern gray squirrel, for example, some melanistic animals have banded hairs, which appear black because the band of phaeomelanin is narrow and the terminal band of eumelanin is broad. These animals have agouti hairs on the abdomen as well, which may appear brown or orange from a distance. Others, like some of the squirrels imported to the National Zoo from Canada in 1902, are completely black, with no banding of the hairs.

When there is a defect in melanin production so that no black pigment is produced, the result is albinism. This is caused by an allele (c) of the color gene (C). The first step in the biochemical pathway for the production of melanin requires tyrosinase, and alleles of the color gene either affect tyrosinase directly or inhibit this first step. In albinism, the pathway is blocked completely at this first step, and no melanin is produced in the body, resulting in white hair and pink eyes. The white "color" results from the reflection of light from the cells containing no melanin. In natural populations albinism is rare for several reasons. First, albinism causes poor eyesight, which is an extreme disadvantage to a squirrel, and second, all-white squirrels are more conspicuous to predators. Albinism is found occasionally in many species, including the eastern gray squirrel, the Eurasian red squirrel, the North American red squirrel, and the eastern chipmunk.

In some towns, people have proudly maintained populations of albino squirrels, usually the eastern gray squirrel. Olney, Illinois, even calls itself

Squirrels: The Animal Answer Guide

A juvenile white eastern gray squirrel. Populations of white, nonalbino squirrels have been maintained in several communities around the United States, most famously in Olney, Illinois. Photo © Center for Biodiversity Studies, Western Kentucky University

"The Home of the White Squirrels." It is not clear whether these white squirrels are albino, because they are reported to have blue eyes. Reputedly, they became established there in 1902. Other genes can also cause white hair, without causing pink eyes and poor eyesight. All-white squirrels with dark eyes are known (e.g., *Callosciurus finlaysonii finlaysonii* in Thailand), but the genetics for this condition are not fully understood in squirrels.

In Washington, D.C., in the vicinity of the National Mall, we see another whitish gray squirrel, which we call "silver." This silver is not the same as the silver gene in the mouse, which causes a mixture of white hairs and dark hairs in the coat. The silver appearance of the eastern gray squirrel is caused instead by long pale tips on the hairs, which are still faintly banded. The genetics of this coat color are unknown, but from specimens in the Smithsonian collection we know that it has existed on the Mall for almost 90 years.

What about patterns of coat color?

Coat patterns differ widely among squirrels, just as they do among other vertebrates. Perhaps the most consistent coat pattern is the difference in colors between the dorsal surface and the ventral surface. In most squirrels, abdomens are white or at least paler than the color of the back. This coat pattern acts as a form of counter-shading that aids in camouflaging the squirrels (discussed in chapter 4, "How do squirrels avoid predators?").

Many species have alternating stripes of pale and dark hairs on the back, like the chipmunks and many ground squirrels of North America (*Tamias* and *Spermophilus*), the palm squirrels of India (*Funambulus*), the rope squirrels of Africa (*Funisciurus*), and some other terrestrial and arboreal squirrels of Southeast Asia (*Tamiops*, *Lariscus*, and *Menetes*). The similarities in striping between palm squirrels and chipmunks are especially striking. We know

of Indian visitors to the northern United States who have been concerned about how the palm squirrels would survive the winter, not knowing that the chipmunks would be comfortably hibernating in their burrows when frigid Canadian air masses affronted us humans. In some species of *Spermophilus*, such as the 13-lined ground squirrel (*Spermophilus tridecemlineatus*) there are both stripes and spots, which appear simply as broken stripes and are probably under similar genetic control.

Other coat patterns in squirrels are common and diverse. Facial markings include spots behind the ears, colored ear tufts, and eye rings. Extremities, such as hands, feet, and tail tips, may differ in color from the arms, legs, or more basal part of the tail. Flash patterns may occur on the arms or hips. Different parts of the trunk or limbs may exhibit different coat colors. All of these patterns are under genetic control, but there are no clear analogies to coat patterns with known genetic bases in other animals. Thus, we can only speculate on how these patterns are genetically determined. What we do know is that, in the embryo of vertebrates, the neural crest cells lying along the spinal cord give rise to the pigment cells—the melanoblasts—which migrate laterally and ventrally. Where the melanoblasts migrate to and how many of them reach a particular area of the skin will partially determine the color pattern, so the genetics controlling their migration in the early development of the embryo are being studied (although not in squirrels). Exciting research and discoveries lie ahead.

Seasonal changes in coat color patterns are obviously caused by other factors, which are controlling the pigment-producing cells. The eastern gray squirrel, for example, has brown agouti hairs on the back and sides in summer, but in the winter the same pigment cells can produce gray agouti hairs, causing the color and pattern in winter and summer to differ.

Squirrels: The Animal Answer Guide

A golden-mantled ground squirrel (*Spermophilus lateralis*) showing the distinctive stripes characteristic of the genus *Tamias*. Photo © National Park Service

Are there age-related differences in coat color?

Some mammals exhibit great differences between the coat color of young individuals and adults. For example, very young baboons and young leaf monkeys have coat colors that contrast with those of adults. Presumably, this is a signal that reduces the aggression of adults toward the young. Squirrels do not have such a distinct juvenile pelage. Perhaps this is because they are raised in nests that commonly are carefully defended by the mother. It is possible to distinguish the young of the year from adults by slight differences in the coat color of the pelage in some species, but coat color differences are trivial compared with those of some other mammals.

Are there seasonal differences in coat color?

There is some seasonal variation in the coat color of squirrels. Take the eastern gray squirrel as an example; summer animals in normal pelage are much browner on the sides, and winter animals are much grayer, with white fur on the back of their ears. In the North American red squirrel, in Minnesota, the winter pelage has a red band down the middle of the back and no black streak separating the ventral and dorsal colors. The summer pelage is darker, lacking the red band, but with a black streak on the side, separating the white abdomen from the darker back. Seasonal differences are less apparent in most other species of north temperate and Arctic squirrels and do not occur among tropical species, except for some sun bleaching of the hair over time.

The coat color of many species of squirrels varies geographically. These two eastern fox squirrels are the same species but notice the black head in the top squirrel, from the southern United States, and the more uniform coloration of the bottom squirrel, characteristic of fox squirrels in the middle and northern parts of the country. Top: Photo © Danielle Munim; bottom: Photo © Gregg Elovich, www.scarysquirrel.org

Is there much geographic variation in squirrel species?

Within-species geographic variations of coat color is very common in squirrels and is commonly the basis on which subspecies are determined. For example, the midwestern fox squirrel (*Sciurus niger rufiventer*) is very reddish, which is why it is called a fox squirrel. On the Delmarva Peninsula to the east of the Chesapeake Bay, the endangered Delmarva fox squirrel (*S. n. cinereus*) is all gray—grayer than an eastern gray squirrel. Further south along the Atlantic Coast, the fox squirrel (*S. n. niger*) commonly has a black head and white nose.

The Eurasian red squirrel (*Sciurus vulgaris*) also exhibits striking within-species coat color variation—from vivid red in some regions to almost all black in others. Their belly fur is always light, though, unlike the all-black forms of the eastern gray squirrel. In some regions of Europe, several color variations may coexist within an area, whereas in other regions, such as in the United Kingdom, only one color morph (the red one) is present. Luc Wauters, of the University of Insburia, and colleagues examined coat color variation in red squirrels in northern Italy. They found some evidence that ecological factors, such as tree type and altitude, were influencing which color morph was most common in an area; more study is needed, however.

Squirrels: The Animal Answer Guide

Eastern gray squirrel (*Sciurus carolinensis*), North America. Photo © Donald Reeve, www.scarysquirrel.org

Eastern fox squirrel (*Sciurus niger*), North America. Photo © Gregg Elovich, www.scarysquirrel.org

North American red squirrel (*Tamiasciurus hudsonicus*), North America. Photo © Plaistow John, www.scarysquirrel.org

Arizona gray squirrel (*Sciurus arizonensis*), North America and Mexico.
Photo © Robert Shantz, www.rshantz.com

Abert's Squirrel (*Sciurus aberti*), North America and Mexico. Photo ©

Robert Shantz, www.rshantz.com

Eurasian red squirrel (*Sciurus vulgaris*), Europe and North Asia. Photo © Gil Wojciech, Polish Forest Research Institute, www.Forestryimages.org

Red-tailed squirrel (*Sciurus granatensis*), Central and South America

**Deppe's squirrel (*Sciurus deppei*),
Mexico and Central America.** Photo ©
Tracey Dixon, www.trp.dundee.ac.uk/~BL

**Northern flying squirrel (*Glaucomys
sabrinus*).** Photo © Phil Myers

**Southern flying squirrel (*Glaucomys
volans*).** Photo © Caroline Thorington

Prevost's squirrel (*Callosciurus prevostii*), Southeast Asia. Photo © Jesse Cohen, National Zoo

Plantain squirrel (*Callosciurus notatus*), Southeast Asia. Photo © Jesse Cohen, National Zoo

Finlayson's squirrel (*Callosciurus finlaysonii*), Southeast Asia. Photo © David Behrens **Smith's bush squirrel (*Paraxerus cepapi*), Africa.** Photo © Jessie Cohen, National Zoo

Green bush squirrel (*Paraxerus poensis*), Africa. Photo © Alain Devez

Forest giant squirrel (*Protoxerus stangeri*), Africa. Photo © Alain Devez

Low's squirrel (*Sundasciurus lowii*), Southeast Asia. Photo © Jesse Cohen, National Zoo

Indian giant squirrel (*Ratufa indica*), India. Photo © Sudhir Shivaram

Pale giant squirrel (*Ratufa affinis*), Southeast Asia. Photo © Tan Chin Tong, www.scarysquirrel.org

One of the most dramatic examples of geographic variation in coat color is found in Finlayson's squirrel (*Callosciurus finlaysonii*) in Thailand, which is all black in some areas, all white in others, and a mix of pelage patterns and coat colors in still other areas. There are less extreme examples of geographic variation in Prevost's squirrel (*Callosciurus prevostii*), which is tricolor in Malaysia, has an all-black subspecies in part of Sumatra, has another black subspecies in Borneo, and is bicolor in other regions of Southeast Asia. More commonly, geographic variation follows environmental factors, and squirrels are darker in wetter habitats and paler in more arid habitats.

Chapter 4

Squirrel Behavior

Are squirrels social?

Yes and no. The social systems of the 278 species of squirrels in the world range from completely asocial to highly social, with many permutations of sociality in between.

Some live solitary lives except when mating; others live in complex social groups and have extensive social interactions with other members of the group.

GROUND SQUIRRELS. Among the most social of the squirrels are certain species of ground squirrels, marmots, and prairie dogs. Gail Michener, of the University of Lethbridge in Alberta, Canada, has been studying ground squirrels for 37 years, and has identified five grades of social organization among the ground squirrels: asocial, single-family kin clusters, female kin clusters with male territoriality, polygynous harems with male dominance, and egalitarian polygynous harems. Woodchucks and Franklin's ground squirrels are classified as asocial, which means they are characterized by solitary living, no amicable interaction with other conspecifics (except mating), and little amicable interaction with offspring. At the other end of the spectrum are the highly social black-tailed prairie dogs and the Olympic marmots.

Yellow-bellied marmots, a frequently studied squirrel, are moderately social, living in polygynous harems with male dominance. These harems are composed of females, yearling young, babies, and a dominant male who defends his territory from potential usurpers. The females in a harem are most often related, and the social relationship among all members of the harem is almost always amicable. The older females in a harem tend to be the only breeding females, with the younger females being reproductively suppressed by the older females' presence. The younger, nonbreeding females are important, though, in that they assist in the care of babies. The dominant male will act aggressively to yearling males, which encourages them to disperse.

South African ground squirrels in southern Africa have been studied extensively by Jane Waterman, of the University of Central Florida, who has documented their social structure. Female South African ground squirrels live in all-female groups, similar to some other ground squirrels, which are made up of only related females and young. These females feed on communal ranges and live communally in several burrows, except during lactation when the female lives alone in a separate burrow. Females interact amicably and there is no identifiable hierarchy among them. The unique aspect of the South African ground squirrel social system are the males. Male South African ground squirrels do not live with females, but instead they live together in groups of up to 20 unrelated individuals of varying ages. Members of a male group share a common home range and forage together throughout this range. Although there is some hierarchy and dominance in these groups, interactions among males are largely amicable, with males grooming each other and sharing burrows.

Among all ground squirrels, social behavior is similar. Some examples of social behavior among ground squirrels include play, grooming, vocalizations, and greeting (such as nose-to-nose or nose-to-mouth contact upon encountering another squirrel). Other social behavior involves group care of the young, such as communal nursing and communal hibernation.

Because of the variation in the level of ground squirrel social systems, they are good research subjects for comparative studies. Daniel Blumstein, of the University of California, Los Angeles, and Kenneth Armitage, of the University of Kansas, did one such study in 1998, which compared life-history tactics across various species of social and nonsocial ground squirrels. In the study, they examined both the costs and benefits of living in a social group. One of the primary benefits of social living is increased survivorship. In fact, the authors state, "No matter how we examined the relationship, more socially complex species had a higher survival rate to age one." Increased survivorship results from a decrease in predation due to a larger number of individuals to detect predators. It also results from an increase in cooperation, through allogrooming (which reduces ectoparasites) and cooperative care of young (to include communal nursing, young females as helpers, and communal hibernation).

Social living comes with a cost. With many animals in a small area, the competition for resources, such as food, mates, and burrows, increases. This

Two South African ground squirrels (*Xerus inauris*) greet each other. Glands around the mouth release scents that are used by many ground squirrels to recognize each other.

Photo © Jane Waterman

competition can lead to increased aggression, increased dominance, reproductive suppression of younger animals, and infanticide. John Hoogland, of the University of Maryland Appalachian Environmental Laboratory, reports that infanticide, mostly by other lactating females, is responsible for the loss of 39% of black-tailed prairie dog litters. Social living, according to Kenneth Armitage, also results in later reproduction, fewer females reproducing, and smaller litter sizes. It also increases the potential for the spread of disease and parasites. One extreme example of this is the quick and complete annihilation of prairie dog colonies by the plague.

As mentioned above, not all ground squirrels are social. Numerous species of ground squirrels are asocial, including chipmunks (*Tamias* spp.), woodchucks (*Marmota monax*), 13-lined ground squirrels (*Spermophilus tridecemlineatus*), and Franklin's ground squirrel (*Spermophilus franklinii*).

One view of how sociality evolves among ground squirrels is that the animals first band together to reduce predation or to better use resources. Social behavior evolves from these groupings. Another view is that, because most species of ground squirrels live in harsh environments and hibernate, there simply is not enough time for the young to successfully disperse in the first year. Therefore, young are retained in a family group for another year or more until they are able to disperse successfully. This necessary reten-

Squirrels: The Animal Answer Guide

Young prairie dogs greet adult. Amicable relationships among family members are strengthened by frequent contact. Photo © Shirley Curtis, www.scarysquirrel.org

tion of young is thought to have led to the evolution of social behavior in the ground squirrels.

TREE SQUIRRELS. Tree squirrels are not considered social animals, although they do exhibit some social behaviors from time to time. One of the main examples of this is communal nesting. Eastern gray squirrel females will occasionally nest communally with close relatives during the winter. Interestingly, female fox squirrels do not do this. Male eastern gray squirrels and male fox squirrels also will nest communally on occasion, but in this case the choice of nesting partners does not depend on relatedness, as it does with female eastern gray squirrels. Eurasian red squirrels, Mexican fox squirrels, Abert's squirrels, and Arizona gray squirrels also have been documented sharing nests with conspecifics. One decidedly asocial squirrel is the North American red squirrel, who defends an exclusive territory and, with the exception of mating, does not interact with other conspecifics.

FLYING SQUIRRELS. Not much is known about the social systems of flying squirrels, because they are nocturnal and difficult to observe. The most studied of the flying squirrels are the North American ones, the southern and northern flying squirrels. Both these species of flying squirrel, similar to tree squirrels, will nest communally. At times, aggregations of flying squirrels can be as large as fifty squirrels in one nest hole. Unlike the communal nesting in tree squirrels, flying squirrel aggregations are composed of both male and female squirrels, and also squirrels of various ages. What we don't know for certain is whether these aggregations consist of squirrels that are related. Work is going on at Peter Weigl's lab at Wake Forest University to examine this question. The hypothesis is that, at least in southern flying squirrels, there is some form of kin recognition and that this plays a role in the decision of a squirrel about which individuals it will nest with. Advances in molecular biology are assisting researchers in trying to tease apart this problem.

Do squirrels fight?

Yes, squirrels do fight. Squirrels may fight in a number of situations, such as when establishing dominance hierarchies, determining order during mating chases, defending a mate or territory, and protecting young. Fights are ordinarily quickly determined and are concluded by chases, which are what we normally observe. The results of fights are easily seen and include lacerated ears, bobbed tails, scars, missing fur, and sometimes even blood from fresh wounds. Fighting is costly in time and energy and can result in serious injury. Therefore, in most instances disputes between squirrels are settled with minimal or no physical interaction. Chasing, aggressive body posturing, and vocalizations are all nonphysical ways squirrels can deter other squirrels.

The North American red squirrel (*Tamiasciurus hudsonicus*) is renowned for its belligerent territorial behavior. Stories abound of its persistent attempts to protect its territory from all sorts of other animals, including other larger squirrels, birds, humans, and even lawn ornaments. Pregnant and lactating female black-tailed prairie dogs can be extremely aggressive, defending nursery burrows and protecting their babies from other females. In 1980, John Hoogland documented a 6-hour-long dispute between two pregnant prairie dogs. And, during breeding season, Hoogland observed many male black-tailed prairie dogs with facial injuries and scars, and in two cases he witnessed fights between males that resulted in the death of the losing animal.

How smart are squirrels?

One should judge animal intelligence not from the perspective of human behavior, but from the perspective of how well the animal is adapted to the demands of its own environment. I see little value in asking how well a crow performs as a human being.

WILLIAM HODOS

Throughout the course of his career Thorington has tagged and monitored eastern gray squirrels in his suburban Maryland neighborhood. One cold rainy morning, he found a small, very wet, young female in his trap. He

Two lactating black-tailed prairie dog females battle with each other. Lactating females are responsible for the majority of infanticides in black-tailed prairie dog communities. Photo © Shirley Curtis, www.scarysquirrel.org

tagged and released her. Even though she remained in his neighborhood for several more years raising her litters, he never again found "Blackie" in his trap. On the other hand, he had a big male, "Old Yellow Beads" whom he easily trapped more than six times. It seems the squirrel rather enjoyed the peanut butter he used as bait. And when he would release him, "Old Yellow Beads" would not run away, but rather walk calmly to the nearest tree, sometimes pausing to take some more peanut butter before he left. Which squirrel was smarter—the one that avoided the trap, or the one that took the extra food?

Animal intellect is a thorny subject, but one that has fascinated researchers for decades. How we should first define, measure, and finally compare animal intelligence across species has been and still is hotly debated. Brain size has been considered by some to be a measure of intelligence. Humans and other primates, for example, have relatively large brains compared with body size. In squirrels, it was hypothesized that social ground squirrels, such as prairie dogs, would have a larger brain compared with body size than nonsocial squirrels—based on the assumption that sociality is complex and would require more brain power. This did not turn out to be the case. What was found, instead, was that tree squirrels tend to have larger brains with respect to body size than ground squirrels do. So why would

tree squirrel brains be bigger? It might have something to do with living in a complex three-dimensional environment. Tree squirrels need to be able to move their bodies quickly and safely up, down, over, and across branches and wires. Controlling all of this complex movement requires a lot of brain power and so would require a larger brain. But, does this mean tree squirrels are smarter than other squirrels?

English novelist and publisher Ford Madox Ford once said, "Genius is memory." The brightest birds are considered to be the corvids (jays and crows), which have very good memories. Scatter-hoarding squirrels have shown that they remember where they bury food, and other squirrels remember the location of a good food source from day to day or year to year—an essential survival skill. Squirrels also memorize specific routes through the trees, typically the quickest way to and from their nest to a prime feeding location.

Learning and memory go hand in hand. To survive, squirrels must learn—learn where to hide when danger threatens, learn what foods are good to eat, learn how to open a nut quickly, learn where to find the best food, learn how to build a sturdy home, learn how to interact with other squirrels. Beyond survival skills, squirrels have also proven that they are adept at learning novel tasks. The BBC special "Daylight Robbery" chronicled wild squirrels as they mastered an increasingly complicated obstacle course to reach food. And many of us watch as our own backyard squirrels "outsmart" all attempts to keep them out of our bird feeders. Is this intelligence or just dogged persistence?

As for the two squirrels mentioned in the beginning, we would say they both are smart, each remembering and learning from their individual experience. Blackie had a rather unpleasant experience in the trap and consequently avoided it. Old Yellow Beads, on the other hand, obtained a tasty treat from the trap and as a result did not avoid it. They both adapted their behavior in response to their individual experience.

Do squirrels play?

Youngsters in play practice a different leap. It looks so like an expression of sheer high spirits, "jumping for joy," that its serious purpose might be overlooked.

BARKALOW AND SHORTEN

Squirrels, like many mammals, engage in play behavior. Play behavior, normally restricted to young squirrels, can be solitary or social and has been documented in both tree and ground squirrels. Solitary play behavior can include running, climbing, jumping/leaping, twisting, tumbling, and play fighting with objects (e.g., twigs, leaves). Social play behavior takes place between two (or more) squirrels and can include play fighting (wrestling, tackling, boxing, chasing) and sexual play (chasing, mounting, play copulation). Social squirrels, such as prairie dogs, sometimes engage in solitary play, and solitary squirrels will engage in social play, usually with littermates. Occasionally in some social, ground-dwelling squirrels, adults, including parents, will engage in play with young.

Play behavior begins at a set stage in a juvenile squirrel's development, decreases as the squirrel ages, and eventually disappears. The timing of the onset and disappearance of play behavior varies between species. For example, play in young yellow-bellied marmots begins approximately 3 days after they emerge from the burrow and continues throughout the summer. In contrast, young woodchucks, the only solitary marmot, play briefly, only for the first day or two after emergence. Belding's ground squirrels begin to play when they emerge from the burrow at around 4 weeks of age. They play vigorously for one to two weeks, but by the third week play behavior all but disappears.

The amount of time each day young squirrels engage in play also varies between species and occasionally between sexes. Juvenile Olympic marmots, when not feeding, spend the majority of their time playing in some form with conspecifics. Young yellow-bellied marmots spend about 40% of their time playing, with males engaging in more play than females, whereas young Columbian ground squirrels devote less than 10% of their time to play.

Why do squirrels (or other animals) play? Some propose that play is a way for a squirrel to practice behaviors and actions that it will use as an adult (e.g., play fighting, sexual play, leaping to avoid predators). Many re-

Two black-tailed prairie dog pups play with each other. Photo © Shirley Curtis, www.scarysquirrelworld.com

searchers suggest that play behavior is essential to a squirrel's physiological development, including neuromuscular control, bone growth, lung capacity, and overall physical stamina and coordination. Others suggest that play behavior, specifically social play, enables a squirrel to establish important relationships, including dominance hierarchies, among siblings and other juveniles.

Scott Nunes, of the University of San Francisco, and colleagues studied play behavior in Belding's ground squirrels. They found that juvenile squirrels that engaged in more social play had increased motor skills. This increase in motor skills was correlated with successful male dispersal in the first year. Additionally, females that engaged in more social play as juveniles, weaned more young as yearlings than those females that engaged in less social play. One wonders whether the more successful females play more, or whether the more playful females are more successful.

Nunes and colleagues also found that squirrels that had a higher percentage body fat played more than those that had less body fat. Well-fed young squirrels play more and increase their motor skills at a greater rate, which can help them avoid predators and win in fights. For females, increased play leads to greater reproductive success.

Do squirrels talk?

Squirrels communicate, but they do not use language the way we do. Despite the literature, like the Doctor Doolittle books, that suggests that animals speak to one another, humans are the only animals that use speech.

The vocalizations of squirrels are distinctive, and you can learn to recognize them and the contexts in which they are given. The alarm calls will probably be the easiest to learn. In our neighborhoods, the eastern gray squirrels are most concerned about cats, and their alarm calls, "chuck, chuck, kwa," frequently alert us to the presence of neighbors' cats before we see them. The intensity and repetition rate probably indicate how upset the calling squirrel is. Social ground squirrels, such as prairie dogs and some marmots, are prolific alarm callers. You only have to approach a colony to hear numerous chirps and trills alerting others to your presence. (More information on alarm calling appears below in "How do squirrels avoid predators?")

Baby squirrels also vocalize, and their high-pitched calls will quickly summon a mother squirrel. At first, baby eastern gray squirrels express discomfort with a squeak vocalization, as early as 3 days of age. By 3 weeks of age, they can emit growls, and by 4 weeks intense discomfort is signaled by a short scream. At thirty to forty days of age, they develop a "muk muk" call, given in anticipation of nursing or feeding. It is like a search call or a contact call. This has been allied to the vocalization given during mating chases by male eastern gray squirrels, also described as "muk, muk, muk" or sneezelike calls, which appear to have an analogous context to the infant vocalization.

Vocalizations given during aggressive interactions between eastern gray squirrels include the growl and the squawk growl, which are very different from alarm calls and the calls given by male squirrels searching for an estrous female.

Many squirrels, even nestlings, have a continuum of calls, expressing degrees of intensity. For example, eastern gray squirrels emit a "kuk" or moan when mildly distressed, and a buzz sound when more intensely distressed. Under extreme distress they will emit a scream. This pattern is common among many of the squirrels that have been studied.

Squirrels also communicate by body signals, but these can be more difficult for us to decipher. Common body signals used by squirrels are flicking of the tail and stamping of the feet. More subtle signals involve body positions, head positions, and ear movements. The orientation of the eyes may also play a role in communication.

Even more subtle yet, for us but perhaps not for the squirrels, are signals communicated by odors. One can surmise that some information is being communicated when squirrels sniff one another—their cheeks, sides, or perineum—but we do not know what messages are being communicated. Similarly, when tree squirrels rub their perineum on a branch or leave a few drops of urine on a branch, we do not know what they're communicating to other squirrels, but presumably some information is being exchanged. Black-tailed prairie dogs often use their perianal scent glands during disputes. The secretions from these glands are very strong and skunklike in odor.

How do squirrels avoid predators?

Squirrels, because of their size, habits, and abundance around the world are an important prey item to all types of animals. Because of this, over evolutionary time, squirrels have evolved multiple means to detect and elude predators. Described below are eight common tactics squirrels use to avoid predators.

CAMOUFLAGE. Blending in with the environment is one of the main ways squirrels avoid being detected by predators, and so it should be expected that the coloration of a squirrel's fur would vary depending on the environment in which it lives. It is well documented among many mammalian species that animals living in more humid environments have darker pelage color than animals living in more xeric, or dry, environments. Many tropical squirrels, such as *Callosciurus melanogaster*, have quite dark, almost black pelage, which certainly keeps them well concealed among leaf shadows and wet bark and soil. Chris Smith, at Kansas State University, documented how the pelage among populations of North American red squirrels (*Tamiasciurus hudsonicus*) varied in accordance with humidity—with squirrels living in coastal, more humid, British Columbia having darker pelage than those living in the drier interior. Desert-dwelling squirrels, such as North American and African ground squirrels, on the other hand, tend to be light colored, which helps them to "disappear" against the sandy soil.

In many squirrels, most noticeably in tree squirrels, the belly (ventral) color is much lighter than the back (dorsal) fur color. This is called counter-shading and is used by numerous mammals, birds, and fish as camouflage. The principle of counter-shading is that the animal is less visible when its best-lighted surfaces are darker and its shadowed surfaces are paler. Abbott Thayer, an artist and naturalist, developed this principle in an 1896 article in the ornithological journal *The Auk*. His many ideas—some outlandish— led to great debate, but they also led to the military's use of counter-shading and camouflage in the First World War.

The types of trees on which a squirrel lives can also influence pelage color. Chris Smith discussed how the grayish dorsal fur of eastern gray squirrels (and some eastern fox squirrels) tends to blend in with the grayish bark of beech, maples, and oaks; whereas the reddish fur of North American red squirrels (*Tamiasciurus*) tends to blend in with the more reddish bark of the many species of conifers throughout their more northern range.

The seasonal molt of some squirrels also may be adaptive in helping squirrels remain concealed. For example, the summer pelage of eastern gray squirrels is brown on the sides and back. The winter pelage is paler, gray on the sides and brown only near the midline of the back. In the summer, these squirrels are well concealed by the shadows and leaves of deciduous trees. In the winter, though, when the leaf cover is gone, the squirrels are more exposed. The lighter winter pelage matches the exposed bark of trees and acts as camouflage for these squirrels.

Some researchers have suggested that squirrels living in areas prone to forest fires have evolved melanism, or dark coloration, as a means of camouflage. Richard Kiltie, at the University of Florida, compared fox squirrels from habitats that burn frequently with those from habitats that do not burn frequently and found that melanistic fox squirrels were more common in frequently burned areas.

SHORT NEST TIMES. Baby squirrels begin life with only a slim chance of survival. Young squirrels lack the defenses and the experience of mature squirrels and therefore are easy targets for savvy predators. Baby squirrels are most vulnerable while they are still in the nest and unable to get away from the snakes, birds, army ants, or even other squirrels that might prey on them. Over the course of time, squirrels in some tropical regions such as Africa and Southeast Asia (where there are significantly more nest predators) have evolved a way to limit nest predation—longer gestation times and shorter nest times. This means babies spend more time inside the mother, are more developed when born, and therefore can leave the nest much earlier than other species.

PHYSICAL AVOIDANCE. Squirrels are gifted athletes in fur coats. Tree squirrels race with ease across branches or rooftops and through labyrinths of vegetation and leap without caution from branch to branch. Equally swift chipmunks disappear before your eyes, flattening themselves out and vanishing through invisible openings. All squirrels also are endowed with sharp claws and ever-growing incisors that encourage predators to make a quick and clean kill. Many mammalogists have scratches and bite marks to evidence the painful results of mishandling a squirrel. The squirrels' skills mean even the most savvy predator sometimes ends up empty handed. Author and scientist Lawrence Wishner once observed a hawk swoop down on a seemingly oblivious chipmunk, talons outstretched, only to come up with nothing—the chipmunk, at the last moment, having dashed to safety.

Chipmunks can run up to speeds of 19 kilometers per hour (kph) (12 miles per hour [mph]), and some ground squirrels are also quite fast, with the golden-mantled ground squirrel (*Spermophilus lateralis*) being clocked at 21.7 kph (13.5 mph). The fastest are the tree squirrels that can run almost 26 kph (16 mph) on a flat surface, and once they reach a tree trunk, their claws grip the surface tightly and allow them to move with ease up, down, and around the trunk. Tree squirrels will flatten themselves against a trunk of a tree and maneuver themselves to the side opposite the predator, in particular, human hunters, continuing to move around the trunk, out of sight, if the curious predator persists in looking for them.

VIGILANCE. Squirrels, like all wild animals, have acute senses, which they use in combination to help them identify and avoid potential predators. Because squirrels live in varying habitats—from open plains to dense forests—the environment plays a significant role in how squirrels puts these senses to use.

Most prairie dogs, marmots, and ground squirrels live in relatively open habitats with little to no vegetative cover. Although this means that predators have little trouble seeing the squirrels foraging in the open, it also means the squirrels are able to spot the predators from a reasonable distance (provided they are paying attention). To "pay attention" squirrels often adopt an alert posture—sitting or standing up on their hind legs, scanning the horizon. Even when foraging, ground squirrels remain vigilant, stopping every few seconds to look about while chewing. Dirt mounds, stumps, rocks, and other elevated patches of earth provide a good view of the surroundings and are common spots for the squirrels to adopt alert postures. The amount of time a squirrel spends in the alert posture or looking about while foraging depends on many interacting factors, including group size, vegetative cover, distance from a burrow, the amount of predators in the area, sex, and time of year. For example, yellow-bellied marmots feeding far from a burrow spend more time looking up than those feeding near a burrow. In contrast, the amount of time Olympic marmots spend looking up does not depend on how close they are to a burrow but instead on how many other marmots are close by. Adult Olympic marmots feeding in a group look up about 15 seconds out of every minute, whereas those feeding alone spend 33 seconds looking up per minute.

As we mentioned above in the answer to "Are squirrels social?" increased protection is afforded to squirrels living in social groups. Regarding vigilance, more animals in an area means more eyes watching out for predators. And, in the end, this means less predation and more offspring that survive to the next generation. It is posited that squirrels first congregated into groups because the combined vigilance of the group members reduced predation. And, from these groupings, social behavior such as alarm calling, kin recognition, and nepotism, evolved.

Tree and flying squirrels live in habitats with significantly more vegetative cover. This type of environment provides many hiding spots, but it also can limit the ability for the squirrel to detect an incoming predator (either from the ground or from the air). Many tree and flying squirrels spend significant time on the ground foraging or storing food. Tree squirrels, while scanning for predators, commonly use the same alert posture seen in ground-dwelling squirrels. While vision plays a large role in predator detection in tree squirrels, sound also plays a significant part. Cracking twigs or shuffling leaves may alert a squirrel to a possible danger obscured by tree trunks or high brush. Upon hearing a strange noise, the squirrel will run up the nearest tree and freeze, listening for more cues.

Fox squirrels have a hearing range of approximately 63–56,000 Hz, which allows them to hear higher pitches than humans and is roughly comparable to the hearing of dogs. Prairie dogs, on the other hand, have a hearing range of approximately 16–26,000 Hz, similar to humans. It is possible that this difference in range of hearing in prairie dogs is an adaptation to a more underground environment, since low frequencies carry better underground and, while above ground, vision is their primary means of detecting predators.

All squirrels have excellent vision and their entire field of view remains in focus. This allows a squirrel, frozen in place, to continue to scan for predators without moving its head. In addition, a squirrel's blind spot (the point where the optic nerve attaches to the back of the eye) is located such that it allows them an uninterrupted view of the sky—extremely important for detecting birds of prey.

NESTS AND BURROWS. If a predator poses an immediate threat, the safest thing for a squirrel to do is hide. Knowing the locations of appropriate hiding spots and the ability to get to them quickly is essential for a squirrel's survival. The most commonly used hiding spots are burrows (for ground-dwelling squirrels) and nests and tree holes (for tree and flying squirrels), but in a real emergency any cover will do. Squirrels choose the appropriate hiding spot based on the type of predator. For example, when pursued by weasels or badgers, Belding's ground squirrels dive into burrows with multiple entrances and exits, but if pursued by an aerial predator, a squirrel will use any burrow, even if it only has one opening. This is because terrestrial predators, such as weasels and badgers, can follow squirrels into a burrow, whereas birds of prey cannot.

Squirrels do not always go into hiding at the first hint of a predator. Running to a nest or hiding in a burrow is energetically costly and time consuming. A subterranean marmot, for example, has no way of knowing if a predator is still in the area and so might stay down for hours before reemerging to forage. This results in a significant loss of feeding time, which for a hibernating animal such as a marmot is limited to begin with. Often squirrels submerge only partially into a burrow, peeking out of the opening to assess the danger. University of Washington professor David Barash, while studying Olympic marmots, saw this amusing scenario: "A

(Left) A Belding's ground squirrel (*Spermophilus beldingi*) mother and her pup stand at alert. The alert posture enables squirrels to better see potential danger. Photo © Gregg Elovich, www.scarysquirrel.org (Right) A round-tailed ground squirrel (*Spermophilus tereticaudus*) surveys its surroundings from the safety of its burrow. Photo © Jim Hughes, www.scarysquirrel.org

coyote elicited an alarm call upon approaching an Olympic marmot colony, whereupon all the marmots ran to burrows. When the coyote rushed the nearest one, it [the marmot] entered its burrow, leaving the coyote scratching furiously outside. After a few minutes another marmot about 7 meters away looked out of its burrow, saw the coyote and called, whereupon the coyote ran to this new burrow, but again to no avail. This was repeated four times, after which a frustrated coyote left the area to the marmots and an amused observer."

The location and construction of a nest is important, both as a hiding place for adult squirrels and as protection for baby squirrels. A tree squirrel nest should be strong enough to withstand shaking and concealed such that it is hard for predators to locate. If a nest is disturbed, the mother squirrel may relocate the young, carrying them one-by-one to another nest.

SOCIALITY. Most species of ground squirrels and prairie dogs and all marmots (except the woodchuck) live in social groups of varying size. Researchers have shown that social living among squirrels offers a range of protection against predators. John Hoogland, a prominent prairie dog researcher, found that prairie dogs living in large groups detected predators more quickly than those living in small groups. He also found that prairie dogs in large groups spent less time scanning for predators than those in small groups. David Barash, in his study of marmots, found the same re-

Squirrels: The Animal Answer Guide

Black-tailed prairie dogs, like these in captivity, are considered one of the most social of the ground squirrels. One of the main benefits of living in a social group is increased protection from predators. Photo © Shirley Curtis, www.scarysquirrel.org

sults, with more social-living marmots, such as Olympic marmots, spending less time alert than solitary living marmots, such as the woodchuck.

Another benefit of social living, commonly called "protection by dilution," is that as group size increases, the probability of any one individual being preyed upon decreases. In some group-living squirrels, individuals work collectively to confront, harass, and discourage certain predators, most commonly those that prey on baby squirrels. This mobbing behavior would not be possible without the close association of other squirrels. Finally, the more squirrels living in a given area, the more burrows there are to use as hiding spots when danger threatens.

ALARM CALLS. Most squirrels vocalize when they are startled, threatened, or detect a potential predator nearby. These alarm calls, as they are known, vary in intensity and, depending on the species, can include chucks, whistles, chips, "kee-yaws," trills, and even ultrasonic calls.

Types of alarm calls. The eastern gray squirrel gives a familiar "chuck" and "chuck, chuck, qwaaa . . ." alarm call, most commonly in response to cats, dogs, or humans. They usually give this call from the safety of a tree branch, looking down at the intruder and flicking their tail up and down. Other squirrels in the area respond to these calls by becoming alert, scanning around, and retreating to the safety of a tree. Calls of the European red squirrel (*Sciurus vulgaris*) are comparable to the eastern gray squirrel, and the fox squirrel gives a combination of similar chucks and chatters.

Eastern chipmunks have three distinct alarm calls: a chip, a chuck, and a trill. The chip and chuck calls are given in response to any detection of a predator, either aerial or terrestrial. These calls usually are given in bouts, which can sometimes last up to 30 minutes. The trill, however, is a short call given when a chipmunk is being pursued by a predator, or escaping into its burrow. While a researcher at the Smithsonian, Lang Elliott commented that he could tell the direction a hawk was flying through the woods by following the chipping of the chipmunks.

North American red squirrels, like eastern chipmunks, have three distinct alarm calls. Unlike chipmunks, though, red squirrels discriminate between aerial and terrestrial predators with their calls. The red squirrel gives a "seet" and "seet-bark" call in response to aerial predators but a sharp bark call in response to terrestrial predators.

Among social squirrels, such as marmots and prairie dogs, some use separate calls for different predators and others do not. California ground squirrels use a multiple note chatter call for terrestrial predators and a high-pitched whistle call for aerial predators. Hoary and Alpine marmots and Uinta, Arctic, Richardson's, and long-tailed ground squirrels also have separate calls for terrestrial and aerial predators. Yellow-bellied marmots have two different alarm calls—the trill and the whistle. They do not use them to discriminate between aerial and terrestrial predators, but instead to communicate the level of risk. For example, Daniel Blumstein and his colleagues found that yellow-bellied marmots whistle longer, louder, and faster when the risk is higher. Similar use of variation in call rate and volume to communicate relative risk has also been found in Olympic and Vancouver marmots.

Three species of Southeast Asian tree squirrels, *Callosciurus*, were studied by Noriko Tamura and colleagues from the Tokyo Metropolitan University. All three species were found to have separate calls for aerial and terrestrial predators. When faced with potential terrestrial predators, these squirrels run up a tree, look down at the predator, flick their tails, stamp their feet, and make continuous staccato barks. When an aerial predator is detected, however, the squirrels remain motionless and emit a single bark or chuckle. If the bird flies very close the squirrel sometimes emits a rattle. Other squirrels hearing these calls freeze in place. These three species of squirrels also have a unique call they direct specifically at snakes. This squeak call elicits other squirrels (even of other species) to come into the area, where they all begin squeaking and mobbing the snake.

Smithsonian researcher Louise Emmons studied the vocalizations of nine species of African tree squirrels in the rain forests of Gabon and found, in general, that these squirrels used two main types of alarm calls: high intensity and low intensity. Low-intensity alarm calls, for the most part, were accompanied by tail and foot movements, in situations not presenting immediate danger. High-intensity alarm calls, however, were used in immediate danger situations and were emitted by a motionless or hidden squirrel.

Purpose of alarm calls. The real question surrounding alarm calls is why squirrels call at all. It seems that by calling, a squirrel makes itself more conspicuous to predators. In fact, during one study Paul Sherman, at Cornell University, recorded that calling Belding's ground squirrels were chased more often by predators, and of the six predations he saw, three of those happened to squirrels who just called.

In social squirrels, alarm calling seems to function primarily as a means of nepotism. Many studies have shown that squirrels with kin nearby call more often than those without kin nearby. In some species such as round-tailed and Belding's ground squirrels, only females with babies or grown young nearby call, whereas in other species, such as black-tailed prairie dogs, males with nearby kin (adult or babies) also call. Some species call when even more distant kin is close by (e.g., cousins). Squirrels commonly give alarm calls in bouts, and it is suspected that these recurring calls serve to maintain vigilance among other group members. Studies have found that some squirrels are able to tell when another squirrel is "crying wolf." Yellow-bellied and steppe marmots and California and Richardson's ground squirrels, have been shown to discriminate between reliable and non-reliable individual alarm callers. This means a squirrel can identify and react differently to the individual that gives an alarm call to every movement in the grass, and the individual that gives an alarm call only to authentic predators.

Similar to ground squirrels, mother tree squirrels call more frequently when they have young nearby. Among adult tree squirrels, however, it is hard to determine whether alarm calling is playing a nepotistic role. Adult tree squirrels do not live together in compact, easily observed groups similar to the social ground squirrels; therefore, determining the exact relationship between tree squirrels in a given area is more difficult. The main function of the continuous chucking and barking alarm calls given by tree squirrels is probably not nepotism but predator deterrence. Squirrels commonly give these calls while observing the predator from the safety of a tree branch, flicking their tail. This style of calling informs the predator that it has been seen and can no longer sneak up on the squirrel. The continuous nature of these calls, sometimes lasting up to an hour, simply reinforces this idea. "Vigilant squirrel here!" "You can't sneak up!" "Vigilant squirrel here!" Other squirrels in the area hear these calls and take appropriate action, but it does not appear that the main purpose of the squirrel's calling is to warn its neighbors.

SNAKE DEFENSE. Instead of fleeing from snakes like they do other predators, many squirrel species approach snakes and harass them. The nature of this harassment is similar among all species of squirrels who do it. They approach the snake on all fours—with their bodies stretched long, tail hair

Snake

South African ground squirrels (*Xerus inauris*), from southern Africa, are just one of the many species of ground squirrels around the world that mob snakes. (Top) Squirrels might mob a snake individually, approaching the snake with a pilo-erected tail. (Bottom) More commonly, squirrels will mob snakes in groups (snake in grass to left of picture, not visible). Photo © Jane Waterman

erected—and make alarm calls. At the same time they flip their tail side to side rapidly. Depending on the risk posed by the snake, squirrels may continue to approach the snake, kick dirt or sand at it, and even attack the snake. The squirrel may harass the snake only a short time, or it may continue vigorously until the snake leaves the area. Commonly, other squirrels in the area hear the alarm calls, congregate around the snake, and participate in the *mobbing*.

To the casual observer snake harassment appears to be a risky, perhaps even foolish, behavior on the part of the squirrels. Research has shown that snake harassment actually serves multiple purposes. First, snakes rely on camouflage and stealth to capture prey. When a squirrel detects a snake, it alerts other squirrels in the area by vocalizing. With other squirrels *on alert*, the snake has little chance of successfully ambushing them. By congregating around the snake and continuing to harass it squirrels are able to drive the snake out of the area.

Second, by approaching the snake, a squirrel is able to see it better and gather important information about it, such as its size and type. California ground squirrels can visually discriminate venomous rattlesnakes from nonvenomous gopher snakes and change their behavior accordingly. This

Squirrels: The Animal Answer Guide

type of information is important, as a large snake poses a greater risk than a small one. And a venomous snake, like a rattlesnake, needs to be approached more cautiously than a nonvenomous one.

Third, by *looming* near a snake and kicking dirt or sand at it, a squirrel can elicit information from the snake. When threatened by a squirrel, rattlesnakes shake their rattle, and the sound of the rattle can tell the squirrel a lot about how dangerous the snake is. A warm snake rattles faster than a cold snake, and a large snake rattles at a lower frequency than a small snake. Studies by Donald Owings and colleagues at the University of California, Davis, have shown that California ground squirrels alter their harassment behavior based on the nature of the rattling sounds, approaching warm, fast-rattling snakes more cautiously than cold, slow-rattling snakes.

Finally, in most species of squirrels that mob, the juveniles and babies, not the adults, are most at risk from snakes. This is because adult squirrels are generally too large for the snakes and the adults of many ground squirrel species have some level of venom resistance to the snakes they encounter. So, in many cases the adult squirrels doing the harassment are at less risk than it first appears.

There is a long evolutionary history linking certain squirrel and snake species. For example, present-day California ground squirrels, Pacific gopher snakes, and Northern Pacific rattlesnakes have coexisted for thousands of years. According to fossil evidence, the ancestor of the California ground squirrel, a species of *Spermophilus* (*Otospermophilus*), coexisted with the ancestors of both gopher and rattlesnakes as far back as the middle Miocene, some 10 million years ago. Rock squirrels (*Spermophilus variegatus*) and Mexican ground squirrels (*Spermophilus mexicanus*) have similarly long evolutionary history with various species of rattlesnakes. One result of this long history is that these ground squirrels have evolved resistance to the toxic venom of the species of rattlesnakes they encounter most frequently. This venom resistance allows the adults to engage snakes offensively with little or no risk, to protect their young. Venom resistance is not consistent within a species but varies between populations depending on the prevalence of snakes in their area.

Naomie Poran, then at the University of California, Davis, and colleagues researched the variation in venom resistance among different populations of California ground squirrels. They found that California ground squirrels from a high-density rattlesnake area were able to neutralize the highest amount of venom, with few side effects, and heal quickly. Whereas the same species of squirrel from a snake-free area in Oregon could not neutralize a similar dose of venom, suffered severe tissue damage, healed slowly, and in some cases died. In another study, Richard Coss and colleagues found that a population of California ground squirrels that colonized a snake-free valley 9,000 years ago has only half the venom resistance of their nearest neighbors in a high-snake area. Another population of the same species living in the San Joaquin River Delta, a snake-free region for more than 60,000 years, has almost no venom resistance.

Another result of the long evolutionary history of the co-occurrence of snakes and squirrels is that squirrel pups from such areas innately recognize snakes. Even squirrels raised in a lab can recognize snakes and behave appropriately.

Chapter 5

Squirrel Ecology

Where do squirrels sleep?

As we mentioned in the first chapter, squirrels live in an extraordinarily diverse range of habitats, from rain forests to deserts and valleys to high mountains. Within these habitats, all squirrels require a safe place to sleep, to give birth to and raise young, and to take shelter from inclement weather and predators.

For many tree squirrels such a place is a nest in a tree or in a hollow of a tree. Leaf nests are usually built at least 4.5 meters (15 feet) up, in a solid location, such as the crook of a branch or against the trunk. Squirrels take the architecture of their nests seriously, using twigs, damp leaves, and moss compressed to create a solid base. They build up from there, creating supports with more twigs and filling in with leaves. These circular nests usually average 46 cm (1.5 feet) in diameter but necessarily are larger for larger species. Nests are lined with soft material, such as stripped bark, moss, fur, or even cotton batting from porch pillows. Leaf nests are quite sturdy, despite their outward appearance, and also are waterproof. Most tree squirrels have multiple nests within their home range—typically a well constructed one that is their primary nest and several other less sturdy ones that are used for midday resting or convenient escape from predators. Nests in tree hollows are the homes of choice for flying squirrels and for many tree squirrels if they are available. Tree hollows provide better protection from the elements than leaf nests and, when lined with soft material, are an ideal place to raise young. Some tree and flying squirrels will occasionally nest in unique locations. The woolly flying squirrel nests in caves, and there have been reports of the North American red squirrel nesting underground.

Most ground squirrels nest in burrows, though the structure and complexity of burrows varies widely across species. Black-tailed prairie dog burrows can be quite large and complex. Although they average 5–10 meters (16–33 feet) long and 2–3 meters (6.5–10 feet) deep, there are records of ones 33 meters (108 feet) long and 5 meters (16 feet) deep. Inside primary burrows are nest chambers, which are lined with grass and average around 30 cm (12 inches) high. These chambers are used for sleeping or for raising young. There are multiple entrances to burrows within a prairie dog coterie, and there are also different types of entrances. All five species of prairie dog create "dome-crater" entrances and entrances with no mound. The entrances with no mounds of dirt outside them are typically found on the edge of their territory and used to escape predators or take temporary shelter. Dome-crater entrances and their mounds are found more centrally and lead to primary burrows. These mounds are used as lookout points, to help prevent the flooding of the burrow by encouraging rainwater run-off and to help promote ventilation. The black-tailed prairie dog and the Mexican prairie dog both create an additional entrance type, called a rim crater, which resembles a small volcano and plays the same role as the dome crater.

The antelope ground squirrel has many burrows distributed across its home range. Females do use a nest burrow, in which they raise young, but otherwise, these squirrels are not selective and will sleep in different burrows on different nights. Because the antelope ground squirrel lives in the desert, these multiple burrows offer convenient shelter during the heat of the day. Idaho ground squirrels have three different burrows: hibernation burrows, nest burrows, and auxiliary burrows. Nest burrows are rather complicated, containing up to thirteen tunnels, eleven openings, and seven chambers. Nests are usually located at the deepest part of the burrow. On the other hand, hibernation burrows consist of a single, steep tunnel ending in a nest chamber. Auxiliary burrows are simple, shallow burrows used primarily for shelter while foraging. Some ground squirrel burrows, such as those of Belding's ground squirrels, are used for only a season, whereas other burrows, like those of the black-tailed prairie dog, can persist for many generations.

The eastern chipmunk (*Tamias striatus*) has burrows that range from simple tunnels to complex systems involving many separate chambers, some for food storage, others for debris, and one for a nest chamber. They may be up to 10 meters (33 feet) in length.

The burrows of the African ground squirrels are similar to those of the North American ones. Burrows of the striped ground squirrel (*Xerus erythropus*) are described as being simple—a central chamber with a few entrances—but complexes of burrows may occur where the density of squir-

A female South African ground squirrel collects brush and other vegetation to take back to her nest.
Photo © Jane Waterman

rels is high. The burrows of the Damara ground squirrel (*Xerus princeps*) are similar, but the openings are usually placed in rocky or gravelly areas and are usually well separated from one another. In the closely related South African ground squirrel (*Xerus inauris*), which nests communally, the burrows are clustered and interconnected.

Do squirrels migrate?

Squirrels do not migrate the way birds, bats, caribou, and wildebeest do, with regular seasonal changes in location. We do not know of any cases of squirrels making regular movements of their home ranges in response to seasonal changes in their habitat, such as the flooding of forests during the wet season or loss of leaves during the dry season. Nevertheless, episodic movements of large numbers of squirrels, in particular, the eastern gray squirrel, are well known, and these fit a more general dictionary definition of the word migration. Such migrations were reported in the eighteenth and nineteenth centuries but may have become less common or less extensive during the twentieth century because of the reduction of forests in the Midwest. The most recent case with which we are familiar occurred in the Mid-Atlantic region in 1990. It was noted by a *Washington Post* reporter, Avis Thomas-Lester, under the title, "Kamikaze Squirrels?" because so many squirrels were being killed on the major highways around Washington, D.C. More widespread migrations occurred in Connecticut and New York in 1933, in western New York in 1935, and throughout the eastern United States in 1968, with hundreds to thousands of squirrels swimming across major rivers and lakes. Similar migrations have been reported for fox squirrels (*Sciurus niger*), the Eurasian red squirrel (*Sciurus vulgaris*), and the North American red squirrel (*Tamiasciurus hudsonicus*). Such episodic migra-

Squirrel Ecology

tions are presumed to be caused by large increases in squirrel populations when there is ample food and good survival, followed abruptly by a failure of the mast crop causing squirrels to search for food in new territory.

How many squirrel species coexist in a forest?

There are fascinating differences in the number of squirrel species that exist in different geographic areas. For example, in Kinabalu National Park (4,343 km²/1,700 mi²) in northern Borneo, there are 13 species of tree squirrels and 10 species of flying squirrels. By contrast, in all of North America, north of the Mexican border, an area more than 4,000 times the area of Kinabalu (approximately 19,460,000 km²/7,556,000 mi²), there are only eight native species of tree squirrels and two species of flying squirrels. Conversely, there are 58 species of ground squirrels in North America, but only five species at Kinabalu.

Although there are a total of 10 tree and flying squirrel species in North America north of Mexico, not all of them coexist in the same area. In North America, as well as in northern Eurasia, only one to three species of tree squirrels and perhaps one species of flying squirrel will coexist in a forest. For example, the North American red squirrel (*Tamiasciurus hudsonicus*) is the common squirrel of the coniferous forests of Canada, where it may co-occur with the northern flying squirrel (*Glaucomys sabrinus*), but further south where there is deciduous forest the red squirrel may co-occur with the eastern gray squirrel (*Sciurus carolinensis*), the fox squirrel (*Sciurus niger*), and the southern flying squirrel (*Glaucomys volans*). The eastern gray squirrel and the fox squirrel coexist over a large part of the eastern United States, from Florida to Michigan, commonly with the southern flying squirrel and without the North American red squirrel. In these same forests, the eastern chipmunk (*Tamias striatus*) may also be common. Although not a tree squirrel, the chipmunk is perfectly capable of climbing to the tops of the trees, which it does to harvest acorns, beechnuts, and the like.

There are no flying squirrels or ground squirrels in South America, but otherwise the pattern is much the same, with one to three species of tree squirrels coexisting in a forest. On Barro Colorado Island in Panama, there is only one species, the red-tailed squirrel (*Sciurus granatensis*), although a second species (*Microsciurus alfari*) formerly occurred on the island. On the drier Pacific side of the Panamanian isthmus, Bill Glanz, at the University of Maine, found the red-tailed squirrel coexisting with the variegated squirrel (*Sciurus variegatoides*). In Venezuela, Smithsonian curator Charles Handley reported four species of squirrels, but only two of them, the yellow-throated squirrel (*Sciurus gilvigularis*) and the northern Amazon red squirrel (*Sciurus igniventris*), occur together, in the lowlands of southern

Venezuela. The other two squirrels occur alone, the Guianan squirrel (*Sciurus aestuans*) in higher elevations in southern Venezuela and the red-tailed squirrel in the north. In the neighboring dry forest of Maraca, Roraima, in Brazil, only the northern Amazon red squirrel occurred. In the western Amazon, it is possible that three or four species may co-occur in a forest, but no ecological studies of this exist. Frequently, the tree squirrels within a South American forest seem to be size-graded. A larger species of *Sciurus* may coexist with a dwarf squirrel (*Microsciurus*) and sometimes also with the pygmy squirrel (*Sciurillus*).

In the tropical forests of Africa and Southeast Asia, the pattern is completely different. Smithsonian researcher Louise Emmons studied squirrels in the M'pala forest in Gabon and found nine species of "tree squirrels" coexisting in the same area. She reported that three of these species were almost completely arboreal—the forest giant squirrel (*Protoxerus stangeri*), the sun squirrel (*Heliosciurus rufobrachium*), and the green-bush squirrel (*Paraxerus poensis*). Four species foraged extensively on the ground: the western palm squirrel (*Epixerus ebii*) foraged widely over an extensive range, but nested in tree holes; the fire-footed rope squirrel (*Funisciurus pyrropus*) foraged within a small range and was the only species that nested in terrestrial burrows; two other species Thomas' rope squirrel (*F. anerythrus*) and the ribboned rope squirrel (*F. lemniscatus*), foraged on the ground and low in the trees and nested in arboreal leaf nests. Lady Burton's rope squirrel (*F. isabella*) foraged little on the ground but extensively in low dense vegetation, and nested above ground. Finally, Emmons observed the African pygmy squirrel (*Myosciurus pumilio*) at almost every level of forest, from the ground to the canopy. Emmons concluded that the African squirrels divided the forest habitats much more than do the American squirrels—by habitat, vegetation type, height in the forest, and choice of foods.

In Southeast Asia, British researchers Kathleen MacKinnon and John Payne, in separate studies, reported a similar pattern of squirrel diversity. At Kuala Lompat, in north-central Malaysia where they worked, there are probably eleven species of diurnal squirrels coexisting in the forest, of which seven are common. Three of the common species are frequently seen in the upper canopy of the forest. Two of these are the giant tree squirrels, the black giant squirrel (*Ratufa bicolor*) and the pale giant squirrel (*Ratufa affinis*), both of which feed extensively on seeds, much less on fruit, and not at all on insects. The third canopy species is Prevost's squirrel (*Callosciurus prevostii*), which feeds more on fruit than on seeds and feeds extensively on insects as well. Another species of *Callosciurus*, the plantain squirrel (*C. notatus*) inhabits lower levels in the forest and feeds on seeds, leaves, and sap or bark almost equally. It relies relatively little on arthropods. The slender-tailed squirrel (*Sundasciurus tenuis*) also occurs at lower levels in the forest and is extensively a bark gleaner and sap feeder. The horse-tailed squirrel (*Sundasciurus hippurus*) is active both on the ground and low in the trees and, on the basis of a small sample size, are thought to feed extensively on seeds. The seventh species, the three-striped ground squirrel (*Lariscus insignis*), is the only terrestrial species and feeds on insects and vegetable matter. Two species of *Callosciurus*, the black-striped squirrel (*C. nigrovittatus*) and the gray-bellied squirrel (*C. caniceps*), and Low's squirrel (*Sundasciurus lowii*) were seen, but not resident, on the study site. The eleventh species, the long-nosed ground squirrel (*Rhinosciurus laticaudatus*), was not present on the site. Though less well studied, flying squirrels also occur at Kuala Lompat, and at least five species probably coexist in the forest there—bringing the total to 16! Thus, in Southeast Asia the diurnal squirrels also have different niches in the forest, differing in the height at which they forage and the foods that they eat, much like the tree squirrels of Africa, but unlike those of the Americas and northern Eurasia.

Why are there so many more species of squirrels in African and Asian forests?

Several factors contribute to the increased diversity of squirrel species in tropical African and Southeast Asian forests. Among the most important of these is the length of time that squirrels have been in each location, the rich diversity of foods available for squirrels in these tropical forests, and the complex geological histories of these areas.

You will remember that South American tropical forests have only a few coexisting squirrel species, in contrast to the many species in the forests of tropical Africa and Southeast Asia. Although the climate and vegetation of the South American forests offer the same benefits of tropical forests elsewhere, they do not have the same abundant diversity of squirrel species. Why? The answer is that there just has not been enough time. Squirrels have been in Africa approximately 19 million years and in Southeast Asia for 21 million years. It was only a very recent 3.5 million years ago, however, that tree squirrels made it into South America. Tree squirrels are the only squirrels that made it across the Panama land bridge into South America and their present diversity consists of only moderately closely related species. Evolution is a slow process. The long time spans squirrels have existed in Africa and South Asia have permitted the evolution of many genera and species of squirrels, but 3.5 million years has not been enough time for South American squirrels to diversify to any great extent. If we could return to a South American tropical forest 10 million years from now, we might see squirrel diversity more similar to what we now see in Africa and South Asia.

Squirrels: The Animal Answer Guide

Another reason for the diversity in tropical forests is that tropical forests are highly productive, containing a vast array of vegetation, which in turn generates an equally vast assortment of fruits, nuts, seeds, and flowers. Although seasonal and annual differences occur in the availability of food in tropical forests, they are not nearly as extreme as in the temperate zone. The abundance and diversity of potential food available in tropical forests means that there is more opportunity for specialization by the squirrels than in a temperate forest. Because food availability changes with the seasons, temperate squirrels cannot specialize in only one kind of food. In the fall, squirrels will feed on abundant nuts and seeds, actively storing extras. During the winter, they feed on these buried nuts and seeds. As spring arrives, squirrels begin feeding on fresh buds and other succulent vegetation. Insects, eggs, leaves, and other abundant items enter the temperate squirrel's diet in the summer. Tropical squirrels, in contrast, can specialize more on specific items because certain foods are continually available in tropical forests.

Tropical forests also provide another abundant food, not found in such diversity in temperate forests: arthropods. The humid climate and abundant leaf litter in tropical forests creates the perfect habitat for spiders, ants, termites, beetles, and a host of other creepy crawlies. These protein-rich food sources are constantly available, and so, in response, several species of tropical squirrels feed on them almost exclusively. Although temperate squirrels do eat insects, insects are not a sufficiently consistent and abundant food source to sustain squirrels throughout the year. Instead, temperate-zone squirrels are opportunistically insectivorous.

Smithsonian researcher Louise Emmons studied nine species of squirrels in the tropical forests of Gabon. Three of the squirrel species she studied were strictly arboreal, rarely, if ever, coming to the ground. They obtained all of their food, mainly fruit, from the upper and mid levels of the forest canopy. Another three species of squirrels fed on the same fruit as the upper-level squirrels but did so on the ground. The reason for this division was that at all times of the year there was fruit available in the canopy and, consequently, dropped fruit available on the ground. At no point would a canopy squirrel need to look for food on the ground and at no point would a ground-feeding squirrel have to venture into the trees to find enough food. Emmons also found that among the three species of arboreal feeders there was a small, a medium, and a large species—all specializing on appropriately sized fruits. The same held true for the ground feeders, with the small species specializing on small fruits and the large species specializing on large fruits.

It is difficult for most of us to appreciate the magnitude of the geological and climatological changes that have occurred during the past 35 million years. These changes are the stage on which evolution plays, however, and have dramatically affected the course of evolution. If geological or climatological changes cause continuous forests to be broken up into isolated blocks, the populations of animals within them are also separated from one another. At this time, the isolated populations can begin to evolve in different ways—perhaps by specializing on different foods or by feeding in different strata of the forests—and in enough time these changes can lead to speciation. Southeast Asia is a good example of this. There are currently a multitude of islands in Southeast Asia; 20,000 years ago at the peak of the glaciations, sea level was 100 meters (328 feet) lower and most of these islands were connected with each other. During the Pleistocene glaciations, as sea levels frequently rose and fell, these islands, their forests, and their mammalian populations (including squirrels) were repeatedly separated and rejoined. These changes gave ample opportunity for speciation, and molecular biology is now enabling us to determine when and where the speciation events occurred.

In tropical Africa, the critical changes were not variations in sea level, but variations in rainfall—between very wet times (pluvial periods) and very dry times (interpluvial periods). These variations in rainfall caused forest blocks to be separated from one another during the dry times and to be rejoined during the wet periods, causing isolation of animals and eventual speciation, as we saw in South Asia. In the temperate zones, the effects of glaciations were markedly different. There, glaciers resulted in large tracts of forest vegetation migrating further south, leaving vast areas of land uninhabitable for forest fauna. The forest fauna either migrated south with the forests or went extinct. During interglacial periods, the forests migrated back north, along with their faunas. There were a few isolated blocks of forest that remained in glacial areas (glacial refugia) but these were not nearly as significant as those in Africa or South Asia.

How do squirrels survive in the desert?

Squirrels live in some of the harshest environments in the world. Just as some squirrels use hibernation, torpor, and an efficient metabolism to survive cold winters (see "How do squirrels survive the winter?"), other squirrels have adopted strategies to survive in very hot and arid environments. In the Sonoran Desert of the western United States, for example, summer air temperatures at midday range from 37 to 45°C (100 to 114°F), with surface soil temperatures (at squirrel height) reaching as high as 60°C (140°F). In the Kalahari semidesert in southern Africa, soil temperatures reach 50 to 60°C (122 to 140°F).

One way for desert-dwelling squirrels to tolerate the harsh desert climate is to spend the entire hot and dry season underground in their bur-

This desert-dwelling round-tailed ground squirrel (*Spermophilus tereticaudus*) avoids the hot sand by climbing up into vegetation. Photo © Jim Hughes, www.scarysquirrel.org

rows, in a version of hibernation known as *aestivation*. Physiologically, these squirrels behave in much the same way as squirrels that hibernate through cold winters—they put on body fat, immerge into a burrow, and lower their heart rate, metabolic rate, and respiration rate. The Mohave ground squirrel (*Spermophilus mohavensis*), which lives in the northwestern Mohave Desert in California, aestivates for 6 to 7 months—immerging at the start of hot weather (July) and emerging in midwinter (January or February). The antelope ground squirrel (*Ammospermophilus leucurus*), which is sympatric with the Mohave ground squirrel, does not aestivate.

Another way desert-dwelling squirrels endure heat is to alter the time they spend above ground foraging. The South African ground squirrel (*Xerus inauris*), for example, is active above ground throughout the day in the wintertime, but during the summer they go above ground only during the relative "coolness" of the early morning or late evening. During very hot or very dry periods, foraging above ground may be unproductive, so the squirrels instead rely on food stored in the burrow. Desert-dwelling squirrels may also alter how they forage when above ground. The round-tailed ground squirrel (*Spermophilus tereticaudus*), which inhabits parts of the Mohave, Yuma, and Colorado deserts in the southwest United States, will forage more in the shade during hot weather and climb up into bushes to avoid contact with the hot sand. The South African ground squirrel feeds on tiptoe, minimizing contact with the sand. The parts of the toes that do come in contact with the sand are protected by thick pads made of keratin.

South African ground squirrels have another tool that protects them from the desert heat—their tails. These squirrels inhabit areas with very little, if any, shade. To compensate for this they use their tails as umbrellas to shade themselves while foraging out in the open. In the early mornings and evenings, when it is "cooler" the squirrels do not hold their tails up.

The South African ground squirrel of southern Africa endures mid-day temperatures well over 37°C. One strategy this squirrel uses to increase foraging time is to turn its back toward the sun and use its tail for shade. This "parasol tail" can re-duce the temperature of the squirrel by up to 5°C. Photo © Jane Waterman

When foraging during the heat of the day, they turn their backs to the sun and hold their tails erect behind them like a shield. A study by Albert Bennett and colleagues at the University of California, Irvine, has shown that this behavior can reduce the temperature of the squirrel up to 5°C (9°F) and can almost double the time it can spend foraging.

Burrows are vital to desert-dwelling squirrels. In addition to being places to give birth, sleep, store food, and escape predators, burrows also act as refuges from the heat. By their design, burrows can maintain a fairly constant, cool temperature regardless of the above-ground temperature. For example, the burrow temperature of a round-tailed ground squirrel only fluctuated between 22 and 25°C (68 and 77°F) during a week when above-ground temperatures ranged greatly from –5 to 39°C (23 to 102°F).

A squirrel that has been out foraging in the heat will return to a shaded spot or a burrow and lie prone across the soil surface, belly flat, arms outstretched, and tail flat on the ground. Sometimes, as in the Mohave ground squirrel, the squirrel will dig into the ground and push its body, head to tail, through the cool soil. Because of the contrast between the hot squirrel and the cool ground, the squirrel can dissipate excess body heat quite quickly.

Many desert-dwelling squirrels also are able to withstand body temperatures much higher than most animals, allowing them to forage longer in the heat of the day. For example, the antelope ground squirrel can withstand body temperatures up to 43.3°C (110°F). Their body temperatures are also very mutable, and during any given day of foraging in the heat and cooling in the shade, a squirrel's body temperature can easily fluctuate 5 to 10 degrees. A few desert-dwelling squirrels, such as Townsend's ground squirrel (*Spermophilus townsendii*), have lower initial body temperatures, which also allows them to spend more time in the open before dangerously overheating. A low resting metabolic rate also seems to be an adaptation by des-

Squirrels: The Animal Answer Guide

ert-dwelling squirrels. San Viljoen, of the University of Pretoria, looked at African tree squirrels from different habitats and found that the species living in hotter and drier environments had lower resting metabolic rates than predicted by size alone.

A fascinating study by Glenn Walsberg and colleagues at Arizona State University demonstrated the importance of a squirrel's hair properties in thermoregulation. The study examined the rock squirrel (*Spermophilus variegatus*), which has two distinct coats—winter and summer—that, although identical in color and outward appearance, have different properties. Walsberg's study showed that rock squirrels with summer coats gained 20% less heat than rock squirrels with winter coats exposed to similar conditions. Many desert-dwelling squirrels, while they have light-colored fur, have dark-tinted skin. Walsberg, in another study, determined that the dark skin of desert squirrels, although it slightly increases heat gain, protects the squirrel by reducing absorption of dangerous ultraviolet radiation.

In the arid climate of the desert, squirrels have limited access to water, and it is therefore essential to their survival to reduce both the amount of water they need to consume and the amount of water they lose. A squirrel loses water to the environment through its urine and feces and through evaporative water loss from the lungs and skin. Desert-dwelling squirrels have adapted several ways of reducing the amount of water they lose. Antelope ground squirrels exemplify these adaptations. First, they have efficient kidneys that can produce extremely concentrated urine. Second, they can reduce their fecal water content up to 43%. Finally, they have fewer mucus-producing cells in their lungs, which decreases the amount of moisture lost to the air during respiration—an almost 50% reduction in water loss compared with non-desert-dwelling squirrels. Because they do not lose large amounts of water, the squirrels do not need to consume as much water to maintain their body weight. Most of the water that antelope ground squirrels obtain is not through drinking but from their food. They maximize their water intake by choosing foods with high water content, such as succulent vegetation, insects, and seeds.

How do squirrels survive the winter?

Squirrels are *endotherms*—animals that maintain a relatively high and constant body temperature regardless of the environment. Endotherms produce their own heat internally, using energy from the food they eat to fuel their metabolism. In contrast, *ectotherms*, such as snakes, cannot produce their own internal heat and depend on the environment to warm them. Because squirrels are endotherms, winter is a very stressful time when cold temperatures and lack of food present significant energetic challenges.

One way squirrels combat these limitations is by reducing body temperature and energetic needs, so that they do not need to feed as often (or at all) during the cold winter months. This strategy is what we call *hibernation*, but not all squirrels hibernate! The only squirrels that hibernate are the Holarctic ground squirrels, including chipmunks, prairie dogs, and all species of marmots. No tree squirrels, flying squirrels, or African ground squirrels hibernate. There are two types of hibernators—obligate and facultative. Obligate hibernators spontaneously enter hibernation in response to a combination of external and internal cues, such as day length and annual physiological rhythms. Facultative hibernators do not spontaneously hibernate but will occasionally enter hibernation if environmental conditions are stressful enough (e.g., drought, extreme cold). The black-tailed prairie dog is an example of a facultative hibernator; the marmots are obligate hibernators. The timing of hibernation varies widely among species. Some species have relatively short hibernation periods, whereas other species spend most of the year in hibernation, up to 9 months! In most cases, the farther north or the higher the elevation where the animals reside, the longer the hibernation period will be. Squirrels of some species, such as the woodchuck, hibernate individually, whereas other squirrels, such as the bobak marmot, hibernate together in a shared burrow—most typically a female and her young, and in some species an adult male.

Before a squirrel enters hibernation, it goes through a period of hyperphagia, or excessive eating, to put on enough body fat to sustain it over the winter. At this time, some squirrels double their weight.

Squirrels do not eat just anything in preparation for hibernation. The types of foods they choose are very important. They require foods that contain certain kinds of fatty acids, because these enable the squirrels to store fat that can be metabolized at low temperatures. It is of no use for them to store fat that they cannot use when their body temperature is low. Once the squirrel has gained enough weight it enters its burrow and seals up the burrow entrance. Inside a chamber in the burrow the squirrel will curl up in a sleeping position, its body temperature will drop and its heart rate, respiration, and metabolism will slow, at times dramatically. The squirrel will remain in this state of suspended animation for several weeks to several months. A hibernating squirrel will periodically arouse. The purpose of these arousals is not fully understood, but we do know that body temperature, heart rate, and respiration increase during these times. Still, the squirrel will not leave the burrow or feed during these arousals. In the spring, these arousal periods become more and more frequent until the squirrel finally emerges from the burrow. The timing of emergence will vary from year to year based on outside conditions such as temperature and snow cover.

Although hibernation is a successful strategy for surviving harsh winters,

Squirrels: The Animal Answer Guide

To prepare for hibernation, most ground squirrels, like this captive black-tailed prairie dog, put on a significant amount of fat, which will sustain them through the long winter. Some ground squirrels even double their weight. Photo © Gregg Elovich, www.scarysquirrel.org

a small squirrel, such as a chipmunk, cannot store enough fat to sustain it through a complete winter season without feeding. Instead chipmunks use a combination strategy of food storage and hibernation. Chipmunks undergo bouts of torpor, when their body temperature drops and they become inactive. However, compared with true hibernators, these torpor bouts are short, lasting only a few hours to at most several days. When not in torpor, chipmunks will feed on food they stored in their burrow during the fall. During mild winter days chipmunks may even leave their burrows and forage above ground.

Holarctic tree squirrels do not hibernate, but instead they remain active on all but the harshest winter days. These squirrels store food during the fall, which they will feed on throughout the lean days of winter. Non-flying tree squirrels are able to remain active because of an unusually efficient metabolism, which is one to four times higher than expected based on their body size. In addition, Holarctic tree squirrels can mobilize the energy from their food approximately 14 times faster than average, which means they can quickly use the energy from their food to fire the furnace inside them.

Holarctic flying squirrels do not have the extremely high metabolism of the tree squirrels and do not hibernate, but they make up for this by using social thermoregulation to keep warm. These nocturnal tree squirrels spend the day sleeping huddled together inside a nest. These communal aggregations in southern flying squirrels (*Glaucomys volans*) can consist of a few to as many as fifty squirrels. By sharing a nest, flying squirrels reduce their overall energetic costs. These flying squirrels, like the tree squirrels, remain active throughout the winter, leaving the nest at night to feed on the food they buried during the fall, though sometimes food is stored inside the nest and fed on communally by the squirrels. Northern flying squirrels

(*Glaucomys sabrinus*) are known to leave their nest for an hour or two, even when the temperature is below –30°C.

What is hibernation?

Torpor is when an animal's body temperature drops below normal level. (In humans normal is about 37°C, or 98.6°F). Torpor can be a drop of only a few degrees (such as what happens at night when we sleep) or it can be a significant drop. *Hibernation* is deep torpor by animals during the winter. (Torpor by animals during the extreme heat of summer is called *aestivation*.)

Deep hibernation occurs when an animal's body temperature drops significantly, to only a few degrees above the ambient temperature, which during the winter can be rather cold. This is a controlled drop in body temperature, not passive ectothermy. How low body temperature drops varies among species, but some drops can be extreme. For example, 13-lined ground squirrels will drop their body temperature to approximately 4–5°C (40–41°F) and long-tailed ground squirrels have been known to drop their body temperature to –1 to –1.2°C (29–30°F).

As body temperature and outside temperature drop, heart rate and respiration slow. A marmot has a resting heart rate of approximately 140 beats per minute when the outside temperature is warm, but this will drop to only 15 beats per minute when the ambient temperature is cold (5–8°C, or 41–46°F). The heart rate of a 13-lined ground squirrel in deep hibernation will drop to only four to six beats per minute. The squirrel's respiration, as it slows, becomes irregular, coming in multiple short breaths and then stopping, sometimes for as long as an hour.

The principle behind hibernation can be summed up nicely by Van't Hoff's Rule, which states that for every 10°C drop in temperature, rates of biochemical reactions (metabolism) slow by half. By dramatically dropping their body temperature hibernating squirrels are significantly reducing their energetic needs. This was illustrated by Kenneth Armitage who found that yellow-bellied marmots who hibernated had an energy savings of 70–80% over marmots who did not hibernate.

Squirrels will not spend the entire winter in deep hibernation. Studies have shown that all squirrels periodically arouse during the winter. During these arousal periods body temperature, heart rate, and respiration all increase within a few hours. (The animal rewarms itself—it is not being warmed up by an outside source.) The arousal periods may help the squirrel assess environmental conditions outside of the burrow, and researchers have noted that these arousal episodes become more frequent as spring approaches. The squirrel, therefore, may be using these arousals to decide when to emerge from the burrow. The arousals may also provide some type of

Squirrels: The Animal Answer Guide

physiological protection. For example, sleep plays an essential role in maintaining good neurological function; however, squirrels do not sleep while in deep hibernation. Researchers have found that squirrels do sleep during the arousal periods. Muscle fitness, cardiac fitness, and gonadal development are other processes that may benefit from periodic arousal. How much time a squirrel spends in torpor during hibernation varies among species. Woodchucks spend 52% of their hibernation time in torpor, whereas yellow-bellied marmots spend up to 87% of their hibernation time in torpor.

Hibernation is not without risks. If there is a drought or food shortage, squirrels may not be able to put on enough fat during the short summer season to sustain them throughout the winter, and as a result may starve to death during hibernation. If the winter goes on too long, even a healthy squirrel may not have enough fat to survive and may die in its burrow. Extremely cold weather may drop the burrow temperature to a point where the squirrel's body temperature drops too low for it to recover.

For hibernating squirrels, the reproductive season is quite short, sometimes only a few weeks in length. In the spring, the squirrels have to balance the need to conserve energy (by remaining in hibernation) with the need to reproduce. Many adult males emerge from the burrow before the snow melts to get a reproductive head start on the females who are still hibernating. With the ground still covered, these males cannot forage and instead continue to use their fat supplies for energy. There may be no food available for up to a month or longer, depending on snow cover. If a male does not have enough reserve left he may starve or freeze to death after emergence. Or he may be too physically compromised to successfully reproduce. In the cases where males and females hibernate together, such as the black-capped marmot (*M. camtschatica*), this does not pose a problem, because the male and female mate before emergence.

Females, although they remain in hibernation longer than males, must not stay in hibernation too long. Staying in hibernation longer means the female has fewer reserves when she finally emerges and begins to feed. This can affect her reproductive ability, as healthier (fatter) females will have healthier babies. Baby squirrels have only a short time in which to feed and grow before the next winter comes. If a female remains underground too long, she will not be able to give birth in time for her resulting litter to put on enough fat to survive the coming winter.

Do squirrels have enemies?

Squirrels are an important part of nature's food web, and their relative abundance, size, and habits make them excellent prey for a multitude of species. Some common squirrel predators include birds of prey, big cats,

(Left) A rare sight: A black rat snake constricts a southern flying squirrel (*Glaucomys volans*), which it most likely captured up in a tree, near the Black River in southern Missouri. Photo © Steve Chervitz (Right) A red fox carries a recently killed ground squirrel in its mouth. Photo © National Park Service

foxes, owls, snakes, martens, bears, wild dogs, wolves, coyotes, badgers, ferrets, herons, and even largemouth bass. Raccoons, crows, and various other omnivores have been observed preying on squirrel nestlings; and domestic dogs and cats, as well as humans also prey on squirrels. In Africa, chimps have been observed preying on ground squirrels, and it is reasonable to assume other primates have occasionally preyed on squirrels or nestlings as well. Swarms of driver or army ants also can prey on squirrels, especially when a nest falls in their path.

Squirrels, because of their significant role as prey, are important in the conservation of endangered and threatened species. The spotted owl, for example, feeds on the northern flying squirrel, and the black-footed ferret preys almost exclusively on black-tailed prairie dogs. Changes in the population of either of these squirrel species will affect the predator. If the predator's population is healthy enough, then it can survive seasonal fluctuations in prey density. But in the case of endangered or threatened species, the populations may be so vulnerable that even slight changes in prey density could seriously harm them.

Squirrels are also, at times, preyed on by their own species. Infanticide is common among social ground squirrels, and cannibalism has been observed in several different squirrel species.

Why do squirrels commit infanticide?

In the "game" of evolution, the goal is to get your genes into the next generation. If you have your own offspring then your sons and daughters

Squirrels: The Animal Answer Guide

each carry 50% of your genes, and you will break even if two of them survive to be adults. Your sisters' and brothers' offspring each carry 25% of your genes. Therefore, you can enhance your contribution to the next generation by assisting your brother or sister in raising their offspring (even if you do not reproduce). This is the background for why natural selection favors assisting your kin, and why anything that might appear altruistic (when dealing with direct kin) is really in your own interest.

In the majority of animals in which infanticide occurs, the act is most often carried out by unrelated individuals, to the advantage of their offspring. For example, male lions, when they take over a new pride, often kill all cubs sired by the previous male. The female lions, now without young, go into heat; the new male mates with them and therefore produces his own offspring. This infanticidal behavior is favored by natural selection, because it causes the females to devote all their reproductive efforts to raising the young of the killer, and his young also face reduced competition from older siblings.

Among black-tailed prairie dogs, however, the majority of infanticides are carried out by close relatives of the victimized mother—most often lactating full- or half-sisters. It seems counterintuitive that a squirrel would kill related offspring, thus limiting the transfer of at least some of her genes to the next generation. When John Hoogland, an expert on black-tailed prairie dogs, studied infanticide in this species, however, he found several possible explanations for how a lactating female might benefit from killing the babies of a close relative.

Prairie dogs live in large colonies, which are subdivided into small units called coteries, which are inhabited by a male, several related females, and their young. All activity, such as feeding and mating, occurs within the coterie, and the coterie is defended from other prairie dogs.

The first potential benefit of infanticide is the removal of future competitors. The more members there are in a coterie, the more competition there is for limited resources. By killing off young, the female squirrel is ensuring greater access to resources for her own young.

Second, in most cases, the female committing infanticide also eats all or some of the babies she kills. Most killers are lactating at the time of killing, and so Hoogland posits that the killer may be gaining important nutrients by eating the young. This increased nutrient level may ensure that her own offspring are well nourished and have a better chance of survival.

Third, mothers who have babies defend an area inside the coterie. No other coterie members may enter this area, which limits their already limited foraging area. When the babies are killed, the mother stops defending her territory, and the killer has more area for foraging.

Another benefit of committing infanticide is that victimized mothers help more. Because they do not have babies to care for anymore, the victimized mothers spend more time defending the coterie territory, looking out for predators, and maintaining burrows. This in turn, decreases the amount of time the killer has to devote to these tasks.

Finally, an infanticidal mother reduces the chances that her own infants will be killed, because most killing is done by lactating females. If their infants are killed, they stop lactating and become less of a threat to the infants of the killer.

Do squirrels get sick?

Yes, squirrels, like humans, get sick and can be infected by a variety of parasites, bacteria, and viruses.

PARASITES. Like most warm-blooded mammals, squirrels are host to a wide range of parasites, including fleas, ticks, mites, lice, roundworms, tapeworms, and bot flies. Each of these parasites can cause varying levels of distress, or even death, in squirrels. Fleas and ticks, besides being a nuisance, also carry and transmit viral and bacterial diseases such as lyme disease and plague. Mange mites cause hair loss, paralysis, and sometimes death. Bot flies lay their eggs under the squirrel's skin. As the larvae grow and develop under the skin, grotesque, oozing "warbles" develop across the squirrel's hide. Bot fly infestation is rarely fatal to the squirrel and once the larvae exit the warble the skin heals.

Sometimes the effect of a parasite might not be immediately apparent. In a simple experiment, Peter Neuhaus of the University of Cambridge, removed ectoparasites (mostly fleas) from some female Columbian ground squirrels and left other females untreated. Treated females gained weight between giving birth and juvenile emergence, whereas untreated females lost weight during that time. Treated females had significantly larger average litter sizes and weaned more young than untreated females. This is a good example of how parasites can subtly affect the fitness and life-history traits of an animal.

Parasitism is widespread in wild animals, and the level of parasitism can be quite high. For example, one study found an average of fifty-eight ectoparasites per individual in a population of white-tailed antelope squirrels in Utah.

BACTERIA. Prairie dogs, chipmunks, rock squirrels, and ground squirrels can carry the *Yersinia pestis* bacteria, commonly known as the plague. The plague bacteria are spread between animals by fleas and most likely were introduced to squirrels through contact with infected rats. The plague is virulent and can wipe out whole populations of ground squirrels. Plague was introduced to North America in 1900 via shipping ports in California and, in less than a decade, established itself in wild rodent populations. Because of the social way of life of prairie dogs, plague has had a particularly devastating effect on them. Dense colonies of prairie dogs have a very low survival rate (less than 1%) after plague moves through. Because plague will survive in the fleas in a burrow for a year after the prairie dogs die, it may be a long time before the colony is reestablished. Survival can be greater in colonies that are small, isolated from other colonies, and not densely populated.

The plague bacteria evolved in Asia, and it is thought that this is why there are no highly social ground squirrels, similar to the prairie dog, in Asia. The fleas that feed on infected animals can transmit the plague bacteria to humans, or humans can become infected by handling the tissues of dead or infected animals. A few humans still die of the plague each year, but thanks to antibiotics and modern medicine, the risk of dying of it is significantly less than during the Middle Ages, when millions of people died. In North America, the plague is most commonly found in populations of ground squirrels in the American Southwest, but the Norway rat (*Rattus norvegicus*) remains the primary carrier of the plague throughout the world.

VIRUSES. All warm-blooded animals are susceptible to the rabies virus, which is most commonly transmitted through the bite of an infected animal. The virus, which is almost always fatal, travels through the nervous system to the brain, where it replicates rapidly. The symptoms of rabies in an animal can include progressive disorientation, agitation, strange behavior, and lack of motor control. Squirrels, including ground squirrels, are rarely found infected and are not considered reservoirs for the virus.

Squirrels are also vulnerable to various pox viruses, including squirrel fibroma virus, monkeypox virus, and parapox virus. The parapox virus has recently been implicated as one of the primary causes of the decline of the native Eurasian red squirrel populations in the United Kingdom. The eastern gray squirrels introduced to the United Kingdom carry the virus but are not adversely affected by it. We guess that the eastern gray squirrels spread the disease to the Eurasian red squirrels, which have no immunity to the virus. Some species of African squirrels carry the monkeypox virus. This virus is related to the smallpox and cowpox viruses and is found in central and west Africa. Bush squirrels and sun squirrels in the genera *Funisciurus* and *Heliosciurus* can be vectors of monkeypox, as can several other African rodents. Monkeypox is transmitted to humans through bites or by direct contact with an infected animal's body fluids (e.g., through hunting and eating). It is contracted most frequently by young children, who catch squirrels and eat them. Although monkeypox is native to Africa, it has been introduced to North America. In 2003, 71 people became infected with monkeypox when a shipment of infected African rodents passed the virus on to prairie dogs that were later sold as pets. In infected humans, the virus, which (in rare cases) can be fatal, causes fever, fatigue, swollen lymph nodes, headache, and rash.

Woodchucks in the eastern United States carry a hepatitis virus that is remarkably similar to the human hepatitis B virus. Hepatitis-infected woodchucks, like infected humans, also commonly develop liver cancer. Because the woodchuck and human versions of the virus are so similar, woodchucks are now being used as subjects in clinical studies designed to discover the link between hepatitis B and liver cancer, and to develop medical treatments for infected humans.

How do squirrels influence vegetation?

Food hoarders can have pervasive influences on the evolutionary biology of the plants they disperse. Some hoarders have acted as strong selective agents on nut and seed characteristics that enhance their attractiveness to foragers. These include nut size, nutritional content, conspicuousness and other traits.

Stephen Vander Wall

Squirrels and their various food plants have coexisted for millions of years. Whereas some squirrels are primarily seed predators, other squirrels, mainly temperate tree and flying squirrels, also are important seed dispersers. In areas where there is a long history of squirrel presence, vegetation has adapted to either promote increased seed dispersal or to reduce catastrophic predation, and sometimes a combination of both.

North American red squirrels and Douglas squirrels (*Tamiasciurus hudsonicus* and *T. douglasii*) are primarily seed predators of conifer trees. Pine cones are removed by the squirrel, either eaten immediately, or taken away and stored in a larder. A larder is a large collection of cones within a squirrel's home range, and the moist environment of the larder typically prevents the cone from drying out, increasing its long-term value as a food source. Rarely do cones germinate within a larder, and if they do, they rarely, if ever, establish successfully. In this case, a conifer tree does not benefit from the squirrels and so would do well to protect its cones from these predators. At all times, the squirrel's goal is to minimize feeding and foraging time and maximize energy consumption, while the tree's goal is to

A North American red squirrel (*Tamiasciurus hudsonicus*) harvests cones from the top of a tree in the White Mountains west of Big Lake, Arizona. Photo © Robert Shantz, www.rshantz.com

minimize predation and maximize reproductive success. These competing goals have resulted in the evolution of a variety of traits.

In a 1995 study in the Rocky Mountains, University of Wyoming professor Craig Benkman found that in areas where tree squirrels (*Tamiasciurus hudsonicus* and *T. douglassii*) were present, the cones of limber pine trees were significantly different from limber pine cones in areas where tree squirrels were absent. In squirrel-populated areas, the limber pine cones had thicker seed coats, more resin, fewer seeds in relation to the size of the cones, and less energy per seed. Presumably, all of these traits have an impact on the ability of a squirrel to use these cones efficiently as a food source. Philip Elliot, of Kansas State University, also looked at the relationship between North American red squirrels and pine cones and found that the presence of these squirrels in lodgepole pine forests supports the development of, "(1) wider cones, (2) a decreasing ratio between the cone length and width, (3) the base of the cone being flush on the limb, (4) cones with a smaller number of viable seeds, (5) a smaller ratio of seed weight to cone weight per cone."

In addition to the structure and content of the cone, how cones are attached to a tree branch also has an impact on the ability of a squirrel to utilize it as a food source. A squirrel must be able to get its upper and lower incisors between the base of the cone and the branch to bite through it. The wideness of the base of the cone, the closeness of the cone to the branch, and the closeness of other cones all influence how effectively a squirrel can do this. Ponderosa pine tree cones, for example, are so difficult to detach from the tree branch that a squirrel must remove an entire branch, take it to the ground, and then remove the cones from there—a very time-consuming process.

Squirrel Ecology

Squirrels, such as the Eurasian red squirrel (*Sciurus vulgaris*), that scatter hoard food are considered important dispersers of the plants whose seeds they bury. The act of removing a seed from under the parent tree and burying it in a more "open" area, either alone or with only a few other seeds, facilitates the germination and establishment of seedlings. Because of this, it is advantageous to trees to attract enough squirrels (and other scatter-hoarding animals) to the crop to ensure distribution but not so many as to encourage overconsumption of the crop by either scatter-hoarding or non-scatter-hoarding animals. To this end, nut-bearing trees have developed several means of ensuring appropriate levels of predation.

First, nuts are attractive to potential seed predators. They are large, contain more energy per unit than any other seed, and fall to the ground in large, easy-to-locate numbers. This advertisement of the nuts, through size and number, by the trees seems counterintuitive at first; but researchers believe these large mast crops are beneficial to the tree. Squirrels continue to scatter hoard food even when they have stored sufficient food to last them through the winter. Because the trees produce more food than the squirrels can actually use in a year, there is a good probability that some of the buried nuts will germinate and become established as seedlings. Many nut trees have years with very large nut crops followed by years of small crops or even no crops. This pattern is erratic and largely influenced by environmental conditions; erratic mast production may benefit the trees, however, because an unpredictable food resource is not as likely to be consistently preyed upon.

Second, nuts are surrounded by tough shells, which protect the nutritious flesh inside. Squirrels are equipped with exceptionally strong jaw muscles and sharp incisors, which can open even the hardest nuts, like the black walnut. If the shells were too hard to open, squirrels could not use them as a food source and would not scatter hoard them. If the shells were easy to open, or did not exist at all, then all predators could gain access, and the crop would be decimated. Therefore, shells serve to limit predation without dissuading animals that act as agents of dispersal, such as squirrels.

Third, nuts have chemical defenses, in the form of tannins, which can protect them from predation. Tannins are a class of chemicals that in quantity make nuts unpalatable and difficult to digest. Studies show that tannins interfere with the ability of some animals to digest protein. Among squirrels, some species are more affected than others. For example, the eastern gray squirrel can successfully digest acorns even with high levels of tannins, but the North American red squirrel would starve to death on a diet of pure acorns because it cannot detoxify those tannins. Different nuts contain different levels of tannins, and tannin concentration can vary within a single nut, with higher tannin levels concentrated around the embryo. It is unclear whether tannins have evolved specifically in response to predation by mammals, or whether tannins evolved for another reason (perhaps as an insecticide or bactericide) and their role as a predator deterrent is secondary. Also unclear is whether certain squirrels have evolved the ability to detoxify tannins as a result of the tannin content in the nuts, or whether that ability existed first and so influenced the type of food the squirrels could exploit.

Chapter 6

Reproduction and Development

How do squirrels reproduce?

REPRODUCTION IN TREE AND FLYING SQUIRRELS. For many tree squirrels and flying squirrels, courtship involves a mating chase. One to many males are attracted to an estrous female by several different cues, including her vocalizations, behavior, and odor. In female eastern gray squirrels estrus lasts only one day. As males congregate near the female, they jockey for the best position—which is as close to her as possible. The dominant male, closest to the female, will cautiously try to approach and mate with her, but she will fend him off. At the same time, the male must watch his back and continue to defend his dominant position from other males. For a while, the situation will be a stalemate. Without warning, the female will race away, and the dominant male and any other males that can keep up will run after her. This is the part of the mating chase that is commonly observed—a female bounding quickly up a tree or across a yard to a hiding place, while a chaos of males tries to follow her or locate her. While hunting for the female, mostly through scent, the males give a characteristic call, described as "muk-muk . . . muk-muk . . ." in the eastern gray squirrel.

The first male to locate the female may mate with her. If another male discovers the mating couple, he will rush in and break up the mating. Mating attacks can be vicious. We observed a male red-tailed squirrel (*Sciurus granatensis*) in Panama break up a mating bout by biting the other male in the scrotum. The female can get hurt as well. During a mating chase of fox squirrels in Florida, we saw a female with a large, fresh gash on her thigh, which probably resulted from such an attack. After the female squirrel mates with one male, the whole mating chase may be repeated. We have observed a female mating with as many as four males during the course of a morning, but others have described a more monogamous situation. By early or midafternoon, the female will become less receptive to mating and the males will lose interest in pursuing her.

For the most part the most dominant males in a chase are the oldest males, but younger squirrels (less than 2.5 years of age) are not excluded from mating. According to studies conducted at the University of Kansas by John Koprowski, young males use an alternative strategy. Young males hang about inside the female's home range but do not actively participate in the chases. When the female breaks away from the chase group and hides, the young males go looking for her. In this situation, says Koprowski, these young males have as good a chance of finding the female and mating with her as the older males.

Other Holarctic tree squirrels, such as the North American red squirrel and the European red squirrel, exhibit the same reproductive behavior. African and Asian tree squirrels are less thoroughly studied, but many species of these have been observed forming mating chases similar to those of Holarctic tree squirrels. Others, like the African sun squirrels, are usually seen in pairs, and they probably have a different, although unstudied, mating system.

The mating chases of flying squirrels probably are very similar to those of diurnal tree squirrels, but because flying squirrels are nocturnal, their mating behavior is harder to observe. Mating chases have been described in some species of Southeast Asian giant flying squirrels, however. In these species the chase begins before dark, and males have been observed doing acrobatic movements while maneuvering for position and access to the female.

Chipmunks, although they are ground squirrels, form mating chases similar to those of tree squirrels. A female begins to vocalize 3 to 5 days before she enters estrus. During this time she sits, usually in a centrally exposed place in her territory, and emits a sharp "chip." Males in and around the area hear the female and start to aggregate near her over the course of several days. The males attempt to mate with her, but she refuses until she enters estrus, which lasts only a few hours. Chipmunks are very territorial, and a female that is not in estrus will aggressively defend her territory. A female in estrus, allows males to enter her territory temporarily, where the mating chase takes place. The actual mating may take place above ground or underground, and the female may mate with several males. During the mating chase males pay no attention to the caches of food that the female has in her burrow, but neighboring females have been known to take advantage of the confusion and steal food from the mating female's larder.

Here is an interesting question about mating chases—Does the female choose the male or males with whom she mates? By vocalizing female chipmunks advertise their position and thus encourage males to aggregate. This, in turn, sets the stage for male competition in the form of mating chases. By being coy, a female squirrel causes the males to form a dominance hierarchy, with the most dominant male closest to her. Presumably, he is the one most likely to mate with her, but she certainly seems able to reject any suitor, if one can judge by the caution with which they approach her. During some chases, it appears as if the male that finds and mates with her is a random choice, but among a chaos of males she may be able to position herself so that the male that finds her is not at all random, but just appears that way to the observer. In his study of North American red squirrels, Chris Smith became convinced that the female decided which male or males she would mate with. Accordingly, it has been suggested that the term *mating chase* is inappropriate, because the female is not being chased but is leading the males on.

REPRODUCTION IN GROUND SQUIRRELS. Ground squirrels have a more varied mating system. In those species that hibernate, the females will go into estrus soon after emerging from hibernation. In some species, the female is in estrus only for a short time, making it easy for a male to defend a female from other males during her entire period of receptivity. In other species, the females have a longer period of estrus, so each female may mate with several males. In some ground squirrels, such as the black-capped marmot (*Marmota camtschatica*) and the bobak marmot (*M. bobak*), the male hibernates with the female(s) and mating takes place underground before emergence from hibernation. For obvious reasons, we know much less about courtship and mating in such animals.

Most marmot species live in social groups that include one adult male and several females. Mating occurs within this social group and is simple, involving nuzzling and nibbling, followed by mounting. The adult male will mate with each female in his group shortly after emergence from hibernation. The hoary marmot in Alaska has only a single female per social group, so in this situation the marmots are monogamous. The woodchuck is the only marmot that is solitary. Within the first 2 weeks after a female woodchuck emerges from hibernation, she will become receptive for mating. A male woodchuck will seek out estrous females. When he finds one, he will cohabit her burrow for a few days and will mate with her. After mating, the male will move on, searching for another receptive female. Thus, the mating system for this marmot species can be considered serial polygamy.

Prairie dogs live in large colonies, which are subdivided into smaller units. In black-tailed prairie dogs, these are called coteries. A coterie commonly consists of an adult male, several adult females, yearlings, and juveniles. The adult females usually mate with the dominant male in the coterie. If there is a second adult male in the coterie, a female may mate with him as well. In some cases, females also will mate with a male from an adjoining coterie. Mating usually takes place underground, perhaps to avoid the interference of other adult males.

African ground squirrels, which do not hibernate, have varied mating systems. The African striped ground squirrel (*Xerus erythropus*) has been observed forming mating chases, but the African unstriped ground squirrel (*X. rutilus*) does not. The South African ground squirrel (*X. inauris*), which lives in separate male and female groups, has an entirely unique mating system. Up to nineteen unrelated male South African ground squirrels form a loosely knit group. Subgroups of males daily check the smaller female groups searching out receptive females, who are asynchronously in estrus multiple times each year. The females are cooperative breeders, jointly caring for their young, a most unusual social system among squirrels.

How long are female squirrels pregnant?

Gestation time, the time from mating to birth, varies between species of squirrels. On average, North American and European tree squirrels have a gestation time of approximately 39 to 44 days. Gestation times in tropical tree squirrels in Africa are longer, averaging between 55 and 65 days, so the young are born more developed. One hypothesis to explain this is that there are more nest predators in Africa, causing natural selection to favor longer gestation and shorter nesting times. In ground squirrels gestation times range from 29 to 31 days in the golden-mantled ground squirrel to 35 days for black-tailed prairie dogs. Many ground squirrels, including all the marmots, hibernate, which means the young, once born, have only a short time to gain their winter weight. Short gestation times in ground squirrels and marmots probably represent a trade-off between the adult female and her young. The female must birth and wean her young in a short enough period to allow her adequate time to fatten up for hibernation, but the young must be mature enough and have sufficient time to grow and fatten up for hibernation themselves. It can be a delicate and critical balance between the competing demands of the mother and her young, particularly because of the variability of weather.

Where do mother squirrels give birth?

All squirrels give birth to young in a nest. Among tree squirrels and flying squirrels, this is usually a nest in a hollow of a tree (in cold weather) or in a leaf nest (in warmer weather). A nest in a hollow of a tree will be lined with leaves, grass, or shredded bark. In the eastern United States, the in-

Three Belding's ground squirrel (*Spermophilus beldingi*) littermates sit at the edge of their burrow. Belding's ground squirrels can have litter sizes ranging from only a single baby up to 10 babies. Photo © Gregg Elovich, www.scarysquirrel.org

ner bark of tulip poplar trees is frequently used for nesting material, and it is common to see the bottommost branches of these trees stripped of their bark. Ground squirrel nests are in a burrow and lined with grass. Among most squirrels, arboreal or terrestrial, pregnant or lactating females may become very territorial and chase other squirrels away from their nest before and after they give birth.

How many babies do squirrels have?

There is a great deal of variation in litter size among squirrels. Tree squirrels in tropical Africa have the smallest litters—only one or two babies per litter—but they are more highly developed at birth. Again, this is suggested to reflect a larger number of nest predators in Africa and selection for rapid maturation. North American tree squirrels average two to four babies per litter. For example, the eastern gray squirrel averages 2.5 babies per litter, whereas the North American red squirrel averages closer to 4 young per litter. North American tree squirrels frequently have more than one litter per year but wean the first litter before they become pregnant the second time. (Author and biologist Fred Barkalow noted that the summer litters of eastern gray squirrels averaged larger than the spring litters during a 10-year study in North Carolina.) On the other hand, certain species of African tree squirrels become pregnant again while still nursing their first litter. Among all squirrels, litter size can vary from year to year depending on the quality and availability of food resources. Litter sizes may increase after a particularly good crop year, and they may be small or nonexistent after a year of poor crops.

Ground squirrels have more young per litter than do tree squirrels of the same weight. North American marmots average 3.4 to 4.2 young per

Table 6.1. Comparison of squirrel gestation and development

	Gestation median age (days)	Lactation median age (days)	Mean age at emergence (days)	Mean age at maturity (months)
Nearctic ground squirrels	29	37.5	31.8	15.21
Nearctic tree squirrels	42	70	51	13.5
African tree squirrels	57.5	50	21.25	N/A
African ground squirrels	46.8	51.25	42	8

Source: Based on Waterman, 1996. N/A = not available.

litter and many species of North American ground squirrels, such as the 13-lined ground squirrel (*Spermophilus tridecemlineatus*), average 6 to 9 young per litter. Because these species hibernate and must put on significant body fat to survive the winter, reproduction is limited to a very short season and only one litter is produced each year. Among the larger ground squirrels and marmots, adult females may breed and produce young only every second year. On the other hand, the South African ground squirrel (*Xerus inauris*) in Africa, which does not hibernate, can have up to three litters per year, but only averages one to two young per litter.

Are all littermates equally related?

Although a single male is capable of fertilizing all of a female's eggs, in many squirrel species the female frequently mates with several males. One result of this multiple mating is that the female can have a litter sired by several fathers. Using DNA testing, Michelle Haynie from Oklahoma State University and colleagues examined the extent of multiply sired litters in Utah and Gunnison's prairie dog populations. They found that 70–90% of all prairie dog litters were multiply sired. Similar percentages of multiply sired litters also have been found in other ground squirrel species (*Spermophilus beldingi, S. beecheyi, S. richardsonii*).

Why would females mate with more than one male? John Hoogland examined this question while studying Gunnison's prairie dogs. He found that the probability of a female Gunnison's prairie dog conceiving and successfully giving birth was 100% when she mated with three or more males. This probability dropped to only 92% if she mated with only one or two males. It is difficult to know whether the female is less likely to conceive because she mates with only one or two males, or whether she mates with only one or two males because she is less likely to conceive, perhaps both being determined by hormones. In a study on Columbian ground squirrels, Jan Murie, professor emeritus at the University of Alberta, suggested that multiple mating by females is a way to enhance the fitness of the offspring by encouraging sperm competition. In theory, when sperm from multiple males compete to fertilize an egg, only the highest quality sperm will be successful.

How long do female squirrels nurse their young?

Lactation time, like gestation time, varies between species. Holarctic tree squirrels, for example, on average nurse their young 70 days, whereas African tree squirrels and African ground squirrels on average nurse their young only 50 days. The probable explanation for the shorter lactation times in African squirrels is the fact that gestation times are so much longer in these squirrels. African squirrel babies are born more developed and as such require less time for growth. The longer gestation times, as mentioned above, are probably an evolutionary adaptation to an increase in nest predators in these regions.

The shortest lactation times, though, belong to the Holarctic ground squirrels, which nurse their young on average only 38 days, with a few species nursing for less than 25 days. Many of these ground squirrels live in very extreme environments and therefore hibernate for a few to several months of every year. Mating, pregnancy, and growth of young must all take place before the next hibernation, which can be only a few short months away.

How fast do squirrels grow?

All squirrels are born altricial, meaning they are born small, with closed eyes, and very limited movement, and some are born with closed ears and no hair. Altricial young require extensive care by the mother to survive. The time it takes for a squirrel to develop from birth to emergence from the nest or burrow varies among squirrel species and is closely related to habitat and environmental conditions.

Smithsonian researcher Louise Emmons has collected development data on several species of African tree squirrels. She found that the green bush squirrel (*Paraxerus poensis*) weighed on average 10% of the adult weight at birth (10.3 grams or 0.4 oz). These squirrels had fine hair and open ears at birth and had open eyes and could run with coordination by 10 days. By 26 days, squirrels could leave the nest on their own, and by 87 days a female squirrel could be pregnant. Another African tree squirrel, Smith's bush squirrel (*Paraxerus cepapi*), developed similarly, opening its eyes at 8 days and leaving the nest on its own at about 19 days. African tree squirrels have a much longer gestation period than other squirrels, and therefore their young are born more developed, reducing the time spent in the nest.

Squirrels: The Animal Answer Guide

(Top, left) A newborn eastern gray squirrel. Photo © Mary Cummins (Top, right) Three-week-old eastern gray squirrel. Photo © Mary Cummins (Bottom, left) Five-week-old eastern gray squirrel. Photo © Mary Cummins (Bottom, right) A baby California ground squirrel (*Spermophilus beecheyi*) illustrates balance and coordination while sitting on its haunches eating. This is an important developmental step that happens between 4 and 5 weeks for Holarctic tree squirrels and earlier for ground squirrels. Photo © Gregg Elovich, www.scarysquirrel.org

Contrast these development data with those of some Holarctic tree squirrels. At birth, most Holarctic tree squirrel infants weigh only 1.8–3% of the adult body weight. These squirrels also have no hair and closed ears. They do not open their eyes until approximately 26–30 days, and it is not until 37–58 days that the squirrels are able to leave the nest on their own. Between 14 and 27 weeks, fox squirrels gain on average 2 grams/day, and the rate is similar for other Holarctic tree squirrels.

Ground squirrel growth rates vary according to whether the species hibernates. Hibernating species have higher growth rates, on average, than nonhibernating species do. For example, the nonhibernating Harris's antelope squirrel (*Ammospermophilus harrisii*) weighs 3.6 grams (0.12 oz) at birth, approximately 2.5% of the adult weight. By 56 days, this squirrel weighs almost 69 grams (2.4 oz) or 50% of adult weight, a 65-gram (2.3-oz) weight gain. In contrast, the hibernating Belding's ground squirrel (*Spermophilus beldingi*) weighs about 12 grams (0.42 oz) at birth, still 2% of the

Reproduction and Development 99

adult weight. In only 42 days, this species weighs 542 grams (1.2 lbs), a gain of 530 grams! Out of necessity, hibernating squirrels must grow quickly in the short season before the next hibernation begins.

The largest squirrels are the marmots, and they have the longest period of growth. This is thought to result from the short summers when food is plentiful, and their long winter hibernation. The Olympic marmots (*Marmota olympus*) do not reach full size and sexual maturity until their fourth summer, when they are 3 years of age. Even woodchucks (*Marmota monax*), which live in a much less stringent environment, continue to grow for 2 to 3 years.

How long do squirrels live?

To determine how long squirrels live, one must be able to identify an individual animal in the wild and follow it throughout its life. Very few of us have the time or resources to identify numerous individuals in a population and follow them for many years, but some dedicated "squirrelologists" have done just this. One of these was Fred Barkalow, who with his students at the University of North Carolina did a long-term study on a nonhunted population of eastern gray squirrels near campus. For 8 years, they captured, marked, and recaptured individual squirrels. From these recapture data, Barkalow was able to compute survival rates, year to year for this population. Amazingly, Barkalow found that approximately 75% of the squirrels in this population died before they reached one year of age, with the average life span of any individual being only 5 months! If a squirrel survived the first year, however, its life expectancy rose significantly. And each year a squirrel survived after that would increase the chances that it would live to see the next year. In this study, females tended to survive longer than males, and in his sample of approximately 1,000 squirrels, only one individual, a female, lived for more than ten years. Tree squirrels in captivity, which are free from many of the pressures that a wild squirrel would face, can live significantly longer, up to 15 years or more.

John Koprowski, at the University of Kansas, did a similar study on eastern fox squirrels in a seven hectare protected urban woodlot and reported that 7 of 321 individuals survived more than 7.5 years. The oldest, a female, lived at least 12.6 years. Two other females lived more than 9.4 years and the oldest male was 8.3 years of age.

Among ground squirrels, the oldest wild woodchuck recorded lived to be 6.5 years old. Yellow-bellied marmots can live up to 14 years, though in the wild it is rare for them to live longer than 8 years. In black-tailed prairie dogs, John Hoogland reported 50–60% survived their first year. During the next 4 years, survivorship of females increased to 70–80% per annum, but stayed closer to 50% for males. The oldest female survived into its ninth year, the oldest males into their sixth year.

Flying squirrels are reputed to live longer than their nongliding counterparts—almost 50% longer on average—but we have not been able to find the data that support these claims. Some researchers believe that the flying squirrels' low metabolism—almost 40% lower than nongliding mammals of comparable size—is responsible for their long lives. Other researchers, such as Donna Holmes and Steven Austad at the University of Idaho, believe that gliding (flight) is responsible for the increased life span of flying squirrels. Other flying animals, such as birds and bats, have considerably longer life spans than nonflying animals of comparable size, even though they have high metabolisms. It is suggested that flight reduces mortality rates for these animals, which, over the course of evolutionary time, has allowed them to evolve delayed reproduction and development and longer life spans. It would be interesting to know if these hypotheses are correct, so perhaps theory will drive field workers to collect the data necessary to test them in flying squirrels, but it will not be easy.

Note that survival rates can vary greatly from one year to the next. In eastern gray squirrels, for example, survivorship during the first year is highly dependent on the autumn seed crop. If there is a good crop of acorns or beech nuts, there may be good survivorship over the winter, which will increase the overall population size. If the nut crops fail in the fall, a very large percentage of the young animals probably will die over the winter.

We must make a similar disclaimer about life span. Any animal's life span, or survivorship, is determined by several interacting factors, many more than can be discussed adequately in this book. These factors include gender, food availability, predation pressure, competition, age at reproduction, litter size, body size, home range size, hibernation, and so on. Thus, life spans may vary among species, but they also can vary between localities and even between years for any particular species. For instance, the failure of a mast crop can have disastrous effects on a cohort of eastern gray squirrels, with very poor survivorship of first-year animals into a second year, and the average life span of this cohort will be greatly reduced.

Chapter 7

Foods and Feeding

What do squirrels eat?

"Practically everything" would be the glib answer to this question. Anyone who has observed squirrels in a backyard or city park knows that the stereotypic image of a squirrel eating only nuts is not the complete picture. Squirrels do not have a reputation for being discriminating and given the opportunity seem to eat almost anything—including candy bars, chicken wings, cookies, and even leftover pizza. In official terms, squirrels are omnivorous, meaning that they eat both vegetable and animal matter. Reflecting this, squirrel species across the world have varied and, sometimes, unique diets.

The foods eaten by squirrels can vary by season, in response to availability. This seasonal availability of food also can vary annually, and it most certainly varies geographically. In the early spring in the temperate zone, tree squirrels (*Sciurus carolinensis, S. niger,* and *S. vulgaris*) and flying squirrels (*Glaucomys volans* and *Pteromys volans*) may feed on a diverse variety of buds, flowers, young shoots, and the like. In the summer, insects can make up a large percentage of their diet. In the autumn, nuts and seeds become the conspicuous source of food. In some species of squirrels, pine cones make up a large proportion of the diet. Abert's squirrel (*Sciurus aberti*) eats primarily seeds from the Ponderosa pine. The North American red squirrel (*Tamiasciurus hudsonicus*) and Douglas' squirrel (*Tamiasciurus douglasii*) also feed primarily on pine cones, most notably western hemlock, spruce, and Douglas fir cones for Douglas' squirrel, and Lodgepole pine and Ponderosa pine cones for the North American red squirrel in the western United States.

California ground squirrels sample a piece of pizza left by a generous human. Squirrels are notorious for some unusual eating habits, taste-testing candy bars, birthday cake, and even chicken wings. Photo © Gregg Elovich, www.scarysquirrel.org

Squirrels do not hesitate to take advantage of opportunity and make use of an abundant resource if it is available. A perfect example of this can be seen during the appearance of the 13- and 17-year cicadas in the United States, when squirrels will gorge themselves, to the exclusion of all other food, on this seemingly infinite food supply. In other cases, squirrels seek out special resources that they need. Chris Smith, at the University of Kansas, observed pregnant and lactating North American red squirrels gnawing on bones and antlers. Presumably these squirrels gain minerals, such as calcium, which is inadequately represented in their normal diet, from the bones, which helps in the growth and development of young. Harmon Weeks and Charles Kirkpatrick looked at sodium (Na) intake in fox squirrels (*Sciurus niger*) and woodchucks (*Marmota monax*). Both species of squirrels are known to lick road salt, and woodchucks are known to eat the salt-laden gravel along the side of the road. Weeks and Kirkpatrick found that both squirrels had peaks in their intake of sodium, with fox squirrels having two peaks (April and September) and woodchucks having one peak (May/June). Sodium is an essential element in the diet, presumably deficient in the diets of these squirrels at this time of year.

Some ground squirrels, such as chipmunks, have diets very similar to those of temperate tree squirrels. Other ground squirrels, such as the marmots, feed extensively on the vegetative parts of plants. Marmots prepare for hibernation by storing energy in their body fat, not by hoarding seeds and nuts. Body fat can only be metabolized when it is liquid, and for hibernators this requires unsaturated fats. Saturated fats are solid at hibernation temperatures and cannot be metabolized. Consequently, marmots and other ground squirrels feed on plants containing unsaturated fats, in

Conifer cones make up a large percentage of the diet of North American red squirrels, like this one in Banff, Canada. Squirrels strip the bracts and eat the seeds underneath. Cones are harvested from the trees, dropped to the ground, and retrieved. Those that are not eaten are stored in an area referred to as a midden. Photo © John Harvey

particular, the fatty acid linoleic acid, which they can metabolize while hibernating and when coming out of hibernation. They carefully select food plants that provide them with the proper amount of these fatty acids—not too much and not too little—to permit their hibernating behavior.

Many species of squirrels feed on mushrooms and other fungi. Small subterranean truffles are a favorite food of northern flying squirrels (*Glaucomys sabrinus*), Abert's squirrels (*Sciurus aberti*), and many fox squirrels. The North American red squirrel (*Tamiasciurus hudsonicus*), Abert's squirrels, the Eurasian red squirrel, and the Japanese squirrel (*Sciurus lis*) will collect mushrooms of many kinds, hang them up on tree limbs to dry, and then return to eat them or carry them away to a larder. You should never trust a squirrel's taste in mushrooms! North American red squirrels will not eat the poisonous Death Angel mushroom (*Amanita phalloides*), but they will collect and eat the poisonous Fly Amanita mushroom (*Amanita muscaria*).

Most tropical species of squirrels have a varied diet, feeding on palm nuts, figs, various seeds, fruits, and insects. Ants, termites, moths, butterflies, beetles, and insect larvae are the most popular insect items. The green bush squirrel (*Paraxerus poensis*) even occasionally eats frogs. Some Asian species of squirrels have quite unexpected dietary preferences. The long-

Many squirrel species change their diet according to the season to take advantage of abundant food supplies. This eastern fox squirrel (*Sciurus niger*) nibbles on buds, an important source of food for many squirrels in springtime because their winter stores are most likely used **up.** Photo © Gregg Elovich, www. scarysquirrel.org

nosed squirrel (*Rhinosciurus laticaudatus*) feeds extensively on ants and termites, and the woolly flying squirrel (*Eupetaurus cinereus*) almost exclusively eats pine needles.

Many species of pygmy tree squirrels (*Nannosciurus, Exilisciurus, Myosciurus, Sciurillus*) are bark gleaners, finding at least some of their food on or under small pieces of bark. What exactly they are getting off of the bark is not well known—it could be small insects, larvae, fungi, or something else entirely. We know of at least one report of squirrels feeding on nectar. Swinhoe's striped squirrel (*Tamiops swinhoei*), from southern China, robs the nectar of a tropical ginger (*Alpinia kwangsiensis*). It bites through the base of the flower to get the nectar, and in the process it may also bite through important flower parts, like the style. Unfortunately, this behavior has dire effects on the plant, because it encourages other nectar robbers, such as ants, discourages the usual pollinators, and directly reduces seed production

Squirrels can be carnivorous. It is not uncommon to see squirrels raiding bird nests for eggs and/or fledglings, and the African sun squirrel (*Heliosciurus rufobrachium*) has been observed capturing and eating adult weaver birds. Some ground squirrel species have been observed preying on baby mammals such as rabbits. Cannibalism also has been observed among some squirrels.

A squirrel's food choice conceals some interesting biology. What and how much a squirrel eats varies with the size of the squirrel. Small squirrels have higher surface area to body volume ratios and therefore lose heat more rapidly to their environment than larger animals. Similar to other mammals, therefore, smaller squirrels run higher metabolic rates than larger ones do. To keep their metabolism high, small squirrels must eat foods that provide them with a quick and rich source of energy—such as insects,

A pale giant squirrel (*Ratufa affinis*) munches on a mango in Ipoh, Malaysia. Photo © Tan Chin Tong, www.scarysquirrel.org

seeds, and nuts. Larger squirrels, such as marmots, on the other hand, are able to eat less nutritious things, such as leaves and other vegetable matter, that need to be fermented in the cecum.

How do squirrels open hard nuts?

When a tree squirrel wants to open a hard nut, like a hickory nut or a palm nut, it first finds a safe place, such as a tree branch, and then starts to work. The rasping noise the squirrel makes as it breaks into the nut can be heard a great distance. The question is, how is the squirrel getting into the nut—with its upper incisors or with its lower incisors? Smithsonian researcher, Louise Emmons, has spent hundreds of hours observing squirrels. She explained that if you watch closely, you can see how a squirrel holds the nut between its hands, braces it with the upper teeth, and rasps with the lower incisors. Thus, the lower incisors do most of the cutting. There seem to be at least two techniques used by different squirrels. The common one is to scrape open a single small hole in the nut and then enlarge the hole. The other has been reported in the Guianan squirrel (*Sciurus aestuans ingrami*) in southern Brazil, which makes three cuts to open a triangular window in the very hard palm nuts on which they feed. The upper incisors of squirrels should not be discounted, because they are more than sharp enough to easily cut through human skin, as many biologists have painfully discovered.

In all cases, nut-opening behavior is learned both through trial and error and observation. Knowing how to open nuts quickly is very important, as the quicker a squirrel can open a nut, the more nuts it can consume in a given period, and the less time it is exposed to predators. Peter Weigl and

Squirrels: The Animal Answer Guide

An eastern gray squirrel in southern Florida attempting to eat a fallen coconut. Photo © Donald Reeve, www.scarysquirrel.org

Elinor Hanson from Wake Forest University explored the difference between trial and error and observational learning. They took 6-month-old captive-bred squirrels and divided them into two groups. Both groups were presented with hazelnuts, which they had never seen before. The squirrels in group 1 were able to watch a mature wild-caught squirrel, caged nearby, eat the hazelnuts, while the squirrels in group 2 were caged in isolation. The researchers then recorded how quickly and efficiently the two groups opened the hazelnuts over a 6-week period.

Weigl and Hanson observed that the squirrels in group 1, which observed the experienced squirrels as they ate, decreased their nut-opening time from 71.6 minutes per nut to 27.8 minutes per nut over the course of 6 weeks. The squirrels in isolation, though, only decreased their opening time from 75.3 minutes per nut to 48.8 minutes per nut. (For comparison, the experienced model squirrel took only 23 minutes per nut.) In addition, the squirrels in group 1 varied little in the technique they used to open the nuts, whereas the squirrels in group 2 varied significantly. These results suggest that observational learning plays an important role in the speed and efficiency with which young squirrels learn to open nuts. Although the isolated squirrels did learn to open nuts, it still took them longer to do so than the observers.

Do all squirrels bury their food?

The short answer to this question is no, not all squirrels bury their food. The storage of food for later use, or hoarding, is extremely common among most squirrel genera, though. Food hoarding is not unique to squirrels—members of 20 arthropod families, 15 bird families, and 29 mammal families hoard food in one manner or another. Animals may hoard to ensure access to food when resources are scarce (during winter or the dry season) and/or when a particular food is abundant (fall acorns, for example). We will use a few commonly known species of squirrels to provide examples of three different strategies of food storage regularly used by squirrels.

The fox squirrel (*Sciurus niger*), eastern gray squirrel (*Sciurus carolinensis*), and the Eurasian red squirrel (*Sciurus vulgaris*) use a form of food storage called scatter hoarding. In this strategy, a squirrel caches individual food items or small collections of food items throughout its home range. The squirrel can cache food in a variety of places, most commonly by burying it shallowly in the ground. Squirrels will also scatter hoard in branch forks, between or under leaves, and in tree cavities. Temperate tree squirrels, which do not hibernate, rely heavily on this scatter-hoarded food during the winter, when other food is scarce. South and Central American tree squirrels, such as the red-tailed squirrel (*S. granatensis*) and the Guianan squirrel (*S. aestuans*), as well as most African tree squirrels and some African ground squirrels also scatter hoard food.

Douglas' squirrels (*Tamiasciurus douglasii*) and North American red squirrels (*Tamiasciurus hudsonicus*), on the other hand, store food in a central location, called a midden, inside their territory. This strategy is called "larder hoarding." Red squirrels will cut off cones or groups of cones from branches and then take them one by one back to the midden. Red squirrels are prolific cachers, and some researchers have estimated they can easily cache hundreds of cones in a day. Middens, which are vigorously defended by the squirrel, grow quite large and often contain enough food for a whole season or two. Not all larder-hoarding squirrels create middens; some other squirrels, such as chipmunks, will store food inside their burrow or nest.

A squirrel can hoard food for very short periods, as Louise Emmons observed in some species of African tree squirrels, which would wedge half-eaten palm nuts in tree branches and retrieve them a few hours later; or for longer periods (e.g., days, weeks, months) as commonly observed in North American tree squirrels.

The final form of food storage is used by the woodchuck (*Marmota monax*) and other marmots. Instead of storing food in a midden or burying it throughout its home range, the woodchuck gains a significant amount of weight, in the form of body fat, before entering hibernation. Strictly speaking, this is not food storage, but it is energy storage and serves the same purpose. The woodchuck will live off of this body fat throughout hibernation, which in some species of marmots and ground squirrels can last up to 8 months.

The evolution of different caching strategies highly depends on the squirrel's ability to protect the cache and the necessity for access to the

North American red squirrels collect conifer cones and store them in a centralized location within their territory. These storage spots, which can sometimes grow amazingly large, are called middens. Photo © National Park Service

cache. A scatter hoarder, for example, by burying food items throughout an area should reduce the likelihood of a single, disastrous raid on its food supply. For this strategy to work, the squirrel must be able to recover more of its buried food than other animals do. To accomplish this, the squirrel must first have a means to "remember" where it buried the food, and, second, it must bury the food in a manner that will deter naïve animals from recovering it. The caches should not be clumped too close together; otherwise a naïve animal coming across one cache could easily find other caches. In addition, the caches should not be too far apart, or the squirrel would have to expend too much energy burying and then recovering the food. Another beneficial aspect of scatter hoarding is that it allows a squirrel to move the location of its nest without the need to relocate a large store of food. This offers the squirrel the flexibility to move a nest at any time because of parasite infestation or season changes.

In contrast, squirrels that larder hoard have a food store that must be protected. Defending such a resource can be energetically costly, so for larder hoarding to be a successful strategy the squirrel must be able to guard the food efficiently. North American red squirrels, for example, have well-defined territories that contain their nest and their midden. They vigorously defend their territory from their own species and other species as well. We used to believe that North American red squirrels were exceedingly successful at defending their middens, but Fritz Gerhardt has demonstrated that pilfering is common. He color marked uneaten cones in middens and

Foods and Feeding

then located them again, finding that ownership changed for 25% of the cones. In all, 97% of the squirrels in his study pilfered cones, some of them stealing far more than they lost to other squirrels. Even so, the advantages of larder hoarding are obvious. By concentrating food within a defended territory near or in a nest, the squirrel is able to guarantee access to food year round. Chipmunks, which store food underground in their burrow, do not even need to go above ground to feed during the winter and instead can feed off their store until the weather improves.

Keep in mind, though, that these food-storage strategies are not mutually exclusive. A squirrel species that larder hoards in one part of its range might scatter hoard in a different part. Variations in hoarding strategies can occur within a population of squirrels, with certain individuals larder hoarding and some scatter hoarding. Finally, individual squirrels can and do employ a combination of strategies. Numerous hibernating ground squirrels that store body fat, for example, also store food in their burrows. This extra food supply may allow them to emerge from hibernation earlier, which can give the males a head start on finding females, or to stay in hibernation longer if the weather is severe.

The particular type of food-storage strategy (or combination of strategies) a squirrel uses is highly dependent on many interacting factors, including, habitat, food type, food quality, food availability, inter- or intraspecies competition, season, age, sex, territory size and location.

How does the squirrel decide what food to store?

One of the immediate decisions that a food-hoarding squirrel must make is the choice of what food it should eat immediately and what food it should store for later use. The sheer range of food items that squirrels encounter and select for eating and caching make the potential outcomes of such choices endless—or so it seems.

STEELE AND KOPROWSKI

The question of how squirrels choose which foods to store has intrigued researchers for many years. Because of several experiments using eastern gray squirrels, we have learned a significant amount about how some of these choices are made.

In the fall and winter, the eastern gray squirrel primarily eats nuts from oak trees—of which two common subgenera occur throughout its range—the white oak and the red oak. White oak and red oak acorns differ in several ways, and these differences have provided researchers with an excellent opportunity to examine some factors that may affect food storage.

The first factor is tannin content. Tannins are chemical compounds found in acorns that, among other things, make them hard to digest and unpalatable. If you bite into a red oak acorn you will find it extremely bitter. Red oak acorns have a high tannin content (6–10% dry weight), whereas white oak acorns have a low tannin content (<2% dry weight). On the basis of observation, squirrels appear to eat more white oak acorns and store more red oak acorns. Leila Hadj-Chikh and Michael Steele at Wilkes University designed a clever experiment to see whether the low tannin content was the reason squirrels ate more white oak acorns. In the experiment, squirrels were offered white oak acorns and red oak acorns that had been experimentally manipulated to have similar tannin contents. If tannin content was playing a role in the squirrel's decision, then an equal number of both types of nuts should have been buried. They found that even with equal tannin content, however, squirrels still cached significantly more red oak acorns and ate more white oak acorns.

Another factor that could influence a squirrel's storage decision is handling time. Large or particularly tough acorns take longer to open and eat than smaller acorns. Squirrels, therefore, should not waste time eating large or hard to open acorns. In another experiment, Hadj-Chikh and Steele found that gray squirrels offered small, easy to open red oak acorns and large, hard to open, white oak acorns still buried more red oak acorns and ate more white oak acorns.

A third factor that could affect a squirrel's storage decision is perishability. Acorns are most valuable to the squirrel in the nut form—when all the nutrients and fats are concentrated. Once an acorn germinates it becomes less nutritious and contains less energy for the squirrel. White oak acorns germinate in autumn soon after falling from the tree, whereas red oak acorns lie dormant throughout the winter before germinating. Because white oak acorns germinate early, it makes sense for a squirrel to eat them right away, because buried acorns germinate quickly, becoming useless to the squirrel later in the season. On the other hand, red oak acorns do not germinate until the spring, so by burying them, the squirrel does not risk losing a valuable food source. This is exactly what Hadj-Chikh, Steele, and other researchers found in various experiments. Squirrels are preferentially burying red oak acorns because they germinate later, thus providing the squirrel with a reliable food source over the winter. This behavior is not exclusive to squirrels. Several other mammal species also preferentially store red oak acorns as well, probably for the same reason. The eastern gray squirrel and the Mexican gray squirrel (*Sciurus aureogaster*) have figured out one way to get around the early germination of white oak acorns. These squirrels have been observed removing the embryo of white oak acorns and then burying them. By doing this they prevent the acorn from germinating. The squirrels do not do this to red oak acorns they bury in the fall, but University of Richmond professor Peter Smallwood along with Michael Steele have observed gray squirrels digging up red oak acorns in the spring just before they germinate, removing the embryos, and then reburying them. Chipmunks have also been observed removing the embryo of beech seeds stored in their larder to prevent them from germinating.

Perishability does not just apply to germination schedule. An acorn infested with insects or disease has no long-term storage value for the squirrel, because it will disintegrate quickly underground. And, in accordance with these findings, squirrels tend to eat infested acorns (either red or white oak) instead of burying them. In fact, insect-infested nuts may offer squirrels an instant nutritional boost, allowing them to continue foraging longer.

North American red squirrels are careful not to store fresh mushrooms in their larder, because these may be host to insect larvae and nematodes. Instead, the squirrel dries the mushrooms by caching them between tree branches. The drying process kills or inactivates the organisms, allowing the squirrel to then safely relocate the mushrooms to its larder.

How do squirrels find the food they have buried?

The scatter-hoarding eastern fox squirrel does not have a centralized food store but instead has hundreds of caches across its home range that it has buried over the course of months. So, when the weather turns bad or fresh food is scarce, how does the fox squirrel locate this stored food? Does it remember, does it use landmarks, or is it just lucky?

It is well documented that certain species of scatter-hoarding birds are able to remember where they store food, but testing this in squirrels has been difficult. Squirrels have a well-developed sense of smell, whereas birds do not. For that reason, determining whether a squirrel is using memory or simply smell to locate its food can be complicated. Other difficulties are also involved in studying cache retrieval in squirrels, especially in the wild. There can be multiple squirrels and hundreds (if not thousands) of cache sites in a given area. Monitoring which squirrel buried which nut, and then determining whether that same squirrel retrieves that same nut days, weeks, or even months later would be a colossal undertaking. Most of the studies on squirrel cache retrieval, therefore, have been done in experimental settings involving a limited number of squirrels in very controlled situations. Even so, they have provided us with valuable information.

Jill Devenport and her colleagues at the University of Central Oklahoma studied cache-recovery behavior of 13-lined ground squirrels (*Spermophilus tridecemlineatus*) and found that they located the majority of the food they cached without using any visual or olfactory cues. University of California, Berkeley professor Lucia Jacobs and her colleague Emily Liman

Squirrels: The Animal Answer Guide

studied cache recovery in eastern gray squirrels and found that the squirrels located more of the food they buried themselves than food buried by other gray squirrels (even if the other food was closer). Isabel MacDonald looked at spatial memory in eastern gray squirrels and found that after 20 days they could accurately locate cache sites within 5 cm more than half of the time. Stephen Vander Wall, of the University of Nevada, did a study with yellow-pine chipmunks (*Tamias amoenus*), and like Jacobs and Liman, he found that the chipmunks located more of the food that they cached themselves than food cached by other chipmunks. He also found that the chipmunks appeared to use landmarks in the environment to help them "remember" the location of the food they buried. Finally, he noticed that chipmunks returned more frequently to caches of large numbers of seeds versus caches of only a few seeds—suggesting that the chipmunks are remembering the contents and the location of the caches they create. In combination, these studies provide us with credible evidence that memory is playing an important role in how squirrels locate their cached food.

The environment may alter how and when memory is used to retrieve food. Studies have shown, for example, that smell is ineffective in finding seeds buried in dry soil, but it is highly effective in wet soil. Therefore, in dry conditions a squirrel may rely more on memory to locate its cache, but in wet conditions it may rely more on smell.

The re-caching of food is very common among scatter hoarders, and squirrels are no exception. Squirrels often dig up food and then rebury it in another location. It has been found that squirrels use re-caching as a way to determine whether a buried item is still good; however some researchers hypothesize that re-caching also is a way for the squirrel to "refresh" its memory on where and what it stored.

Chapter 8

Squirrels and Humans

Do squirrels make good pets?

Author and biologist Fred Barkalow wrote the following about gray squirrels: "If you can ignore having your furniture chewed to bits, the draperies shredded, and droppings scattered willy-nilly about the house, the gray squirrel is the pet for you."

As our selection of this quote suggests, we think that squirrels do not make good pets. We prefer animals that can be housebroken, do not bite when patted, and are not destructive. If you search the Internet, you will find many persons who disagree with us, and you will find instructions for maintaining squirrels in captivity. Before seriously considering a squirrel pet, you should check the relevant rules and regulations where you live. In the United States, wildlife is the property of the state, and state laws frequently forbid the maintenance of wild species as pets without state permits. In some regions, there are also health considerations because of diseases that can be transmitted by squirrels. Prairie dogs, for example, have been implicated in the transmission of tularemia and monkeypox.

Should people feed squirrels?

We recommend putting food out for squirrels, because it encourages careful observation of animal behavior and leads to an appreciation of their natural history. We recommend against feeding squirrels by hand, knowing how quick they are and how sharp their teeth are.

Many people who feed birds also feed squirrels, whether they want to or not. Watching squirrels carefully can be very entertaining and at least as

Some squirrels in urban areas will acclimate to humans, and will even learn to take food directly from a hand. But, given the sharpness of their incisors, hand feeding can be a risky undertaking. Photo © Gregg Elovich, www.scarysquirrel.org

educational as bird-watching. One of our colleagues, Clyde Jones, has even bragged about having a bird-proof squirrel feeder. Attracting squirrels by feeding them can have disadvantages, if they also enjoy feeding on your prized plants or nesting where you do not want them, but generally it seems like an innocuous practice.

Do squirrels feel pain?

Yes, surely they do. Pain is a very important sensation that permits an animal to recognize and avoid dangerous stimuli. An animal that cannot feel pain is at risk, because it will not know it is hurting itself on something sharp, something hot, or something in other ways harmful. It is unlikely that squirrels feel pain the same way we do, however. Some painful stimuli are probably more objectionable to a squirrel than they are to us, and others are less so. Squirrels have different sensitivities for hearing, taste, sight, and touch than we do, so the same is probably true for pain. If you watch squirrels carefully, trying to see the world as they do, you will see the differences.

What do I do if I find an injured or orphaned squirrel?

If you find an orphaned baby squirrel, the best advice we can provide is to get in touch with a wildlife rehabilitator. In the United States there is a National Wildlife Rehabilitators Association and directories of wildlife rehabilitators, including an international directory at the University of Minnesota; all are available on the Internet. Rehabilitators have extensive

A baby eastern gray squirrel is hand fed by a volunteer at Squirrel Rescue in Los Angeles, California. Photo © Mary Cummins

experience with injured and orphaned animals and can be extremely helpful. They also have knowledge of local and state wildlife laws and will hold the appropriate permits that allow them to capture, hold, and treat wild animals.

If you see an adult squirrel acting peculiarly, perhaps dazed by a fall or incapacitated by disease or injury, you should leave it alone. In less dire cases, the squirrel may stand a good chance of recovering on its own. In more severe cases, try to locate an animal control officer. The Internet is again a good source of information. You will probably find that your local government has an Animal Control Office with licensed staff familiar with dealing with wildlife problems. You can also check with local veterinarians, in the expectation that they will know who can deal with the problem. You should be alert for problems potentially involving human health. Ground squirrels and prairie dogs, in particular, can carry nasty diseases, such as bubonic plague, and should be treated circumspectly. You should expect any wild squirrel to harbor fleas and mites, and our experience is that squirrel fleas have a bite that you will long remember. Again, licensed wildlife rehabilitators can be good resources for information on how to deal with injured wildlife.

How can I become a better observer of squirrels?

Becoming a good observer of squirrels in the wild takes time and effort, and the best way to begin is to start thinking like a squirrel. The first and

Squirrels: The Animal Answer Guide

most important step is to learn to recognize individuals. When one squirrel sees another, it sees a recognizable individual, and with practice, so will you. To do this, focus on recognizing the differences between males and females, adults and juveniles. Adult females can sometimes be recognized by their nipples, especially if they have recently been lactating. Most adult male squirrels have a pendulous scrotum. Juveniles and adults can most often be recognized by size and by the playful behavior of the juveniles. Also, when observing, pay close attention to differences in coat color and pattern, scars, injuries, and any other unique features an individual squirrel may have. It may help if you can obtain good photographs of the individuals, which you can compare side by side.

Like us, squirrels are creatures of habit, so you can frequently recognize individuals by their behavior. For example, a squirrel commonly takes the same route when going from one place to another. These familiar routes may be crucial for escape from more dominant individuals or from predators. Similarly, the "home range" of an individual squirrel is frequently distinctive and important. A squirrel commonly seen in the same place is likely to be the same individual. Exceptions to this occur in places where many squirrels are attracted, such as bird feeders, or when the weather is cold and the squirrels are sharing dens. Of course, you do not want to depend on location to identify individuals, because you want to be able to recognize when an individual is out of place—like a chipmunk raiding the larder of a neighbor.

It is also important to observe and learn about the squirrel's environment. If you watch closely, your squirrels will teach you which kinds of plants and animals are their most important foods at different times of the year. Plant and insect guides are available, online and in stores, to help you identify the flora and fauna in your backyard or study area. Observe closely and you even will learn which individual trees or plants the squirrels prefer. Probably not all red oaks or other food plants in your area are equally palatable. Watch to see which plants your squirrels frequent the most. Also, locate the burrows of ground squirrels or look up in your trees for the nests and den holes of tree squirrels and, in the evening, watch where your squirrels go.

Setting up a feeding station in your yard is a great way to observe the social interactions between squirrels, because it encourages many squirrels to congregate in an area. Once you are able to recognize individuals, you can identify which squirrels are dominant and which squirrels are submissive. You can observe whether larger squirrels are always dominant, or whether males or females tend to win in confrontations. Over time, you can also observe how juvenile squirrels interact with and learn from aggressive encounters.

One step to becoming a better observer of squirrels is learning to recognize individuals. This eastern gray squirrel is a female, and you can tell by looking at her nipples that she is or recently has been lactating. Also notice how effective her rear claws are for grappling a tree. Photo ©
Donald Reeve, www.scarysquirrel.org

Your best tool in squirrel observation is an enquiring mind—keep asking yourself, "Why did the squirrel do that?" And do not be disheartened when you cannot answer the question. Remember, a squirrel's senses are different from ours. They have better eyesight, hearing, and sense of smell than we do, so it can be difficult to determine how they are interpreting the world around them. Just keep watching, and you may find out!

How do I know whether I have flying squirrels in my backyard?

In the United States and Canada, you might live in the range of the northern flying squirrel (*Glaucomys sabrinus*) or the southern flying squirrel (*Glaucomys volans*). If you live in parts of Russia, Finland, Northern China, Mongolia, and Korea you may have the Siberian flying squirrel (*Pteromys volans*). If you live in Japan, you may also have the Siberian flying squirrel, the Japanese flying squirrel (*Pteromys momonga*), or even the Japanese giant flying squirrel (*Petaurista leucogenys*). Finally, if you live in or near the right habitat in southern or Southeast Asia, chances are that you have multiple species of flying squirrels.

In much of the United States and Canada, if you have tree squirrels during the day in your backyard, there is the chance that you have flying squirrels there at night. First, check in a field guide to see if you live within the geographic range of either of the two species. If so, in particular, if you have tall mature trees in or near your backyard, it is worthwhile to look for evidence of their presence. An open bird feeder, attached to a tree, will serve as an excellent squirrel attractant. In most areas, you can place nuts such as

Squirrels: The Animal Answer Guide

A common inhabitant of many backyards in the eastern United States, but rarely seen, this southern flying squirrel (*Glaucomys volans*) makes a nocturnal visit to Thorington's birdfeeder. Photo © Caroline Thorington

hazelnuts, which are too large for local birds, on the bird feeder. The tree squirrels will take these during the day, of course, but if you place some on the feeder in the early evening and check it in the early morning, you can easily determine whether you have nocturnal visitors.

If nuts are disappearing overnight, you can watch to see what animal is removing them. A red light may help you see, without bothering the animals very much. One "professional" squirrel watcher, Vagn Flyger of the University of Maryland, wired his feeder so that a bell rang in his living room when the flying squirrels arrived. If your nocturnal visitors are evasive, and you cannot see them, you may still be able to find signs of them, such as gnawed nuts or footprints. Properly placed, an ink pad and a piece of paper or index card could record the paw prints that can help you to determine whether mice or squirrels are making off with your hazelnuts.

Why are squirrels important?

The last word in ignorance is the man who says of an animal or a plant, "What good is it?"

ALDO LEOPOLD

Squirrels, by their diverse habits and abundant distribution, are important actors in Mother Nature's drama, and so when someone asks us "Why are squirrels important?" we scarcely know where to start. We have an impressive list of answers to present to them.

First, squirrels are an important link in the ecological food chain—a prey item worldwide. From hawks to weasels to snakes, squirrels fill the stomachs of a variety of hungry predators, including humans. For some predators, squirrels are their primary food. For example, a study on Goshawks in Sweden found that Eurasian red squirrels made up 86% of the winter diet of female hawks. Arctic ground squirrels are the major food of Alaskan wolverines in summertime, and the wolverines continue to feed on cached bodies of the squirrels throughout the winter. Most important from our viewpoint (not the squirrel's), however, is the role that squirrels play as prey for threatened and endangered species. Northern flying squirrels (*Glaucomys sabrinus*) are the main food item of the endangered northern spotted owl, and black-tailed prairie dogs are the primary food of the vanishing black-footed ferret. In Asia, the Himalayan marmot is an important part of the diet of the rare snow leopard. As a result, the survival of squirrel populations and squirrel habitat is essential for the continued existence of endangered predators.

In 1999, Natasha Kotiliar and her colleagues from the United States Geological Survey (USGS) examined the role of prairie dogs in the Great Plains ecosystem and confirmed what other researchers had proposed—that prairie dogs act as a keystone species in the Great Plains ecosystem. Keystone species are species that play such a significant role in an ecological community that if the species were to go extinct, the negative effect on the remaining community would be disproportionately large in relation to the species' abundance. Prairie dog populations have declined dramatically in the past century because of large-scale poisoning by farmers and ranchers and outbreaks of the plague. It is estimated that prairie dogs now occupy less than 5% of their original range. Accordingly, it is essential to understand how prairie dogs interact with and influence other species. Nine vertebrate species in Kotiliar's study depend heavily on the existence of prairie dogs. Mountain plovers and horned larks nest and feed in the disturbed vegetation surrounding prairie dog colonies; burrowing owls, deer mice, and grasshopper mice utilize prairie dog burrows; and black-footed ferrets, ferruginous hawks, golden eagles, and swift foxes prey on the prairie dogs. Black-footed ferrets (*Mustela nigripes*) are obligate prairie dog predators, which means that without prairie dogs the black-footed ferret cannot survive. Because of the severe decline and in some places extirpation of prairie dogs, black-footed ferrets are almost extinct in the wild. The good news is that there are captive breeding programs working towards reestablishing wild populations, but these re-introductions cannot be successful without viable prairie dog populations to serve as a prey base.

The nine species listed by Kotiliar as heavily dependent on prairie dogs are not the only species that interact with prairie dogs. Many other species, to a lesser extent, prey on prairie dogs, graze or nest on vegetation surrounding the colonies, or inhabit prairie dog burrows. For example, lesser earless lizards are more abundant on black-tailed prairie dog colonies than off of them. It is suspected that the lizards use the prairie dog burrows to hide from predators and to escape extreme temperatures. As well, the lizards seem to "enjoy" basking in the sun on the dirt mounds outside of the prairie dog burrows.

Burrows of other ground-dwelling squirrels also support a variety of species. A live-trap study of 94 woodchuck burrows in Indiana collected (in addition to 35 woodchucks) 20 opossums, 19 cottontail rabbits, 8 raccoons, 2 foxes, 104 white-footed mice, 32 house mice, 10 short-tailed shrews, 2 meadow jumping mice, 2 meadow voles, and 2 masked shrews. Studies on insects found in and around ground squirrel burrows and prairie dog colonies are limited, but it is assumed that insect abundance and insect species assemblages also are affected by these squirrels.

Plants also are influenced by the actions of prairie dogs. As they dig burrows, prairie dogs bring mineral-rich subsoil to the surface. This soil creates an ideal environment for the establishment of plants not able to compete with common prairie grasses. Fetid marigold, scarlet mallow, black nightshade, and pigweed are a few plants mentioned by John A. King that exist exclusively around prairie dog burrows. In addition, prairie dogs remove tall grasses around their burrows, which also assists in the establishment of novel plant species. The urine and feces of prairie dogs contribute nutrients to the soil, which may enhance the growth and decomposition of surrounding vegetation, and also prairie dog burrows enhance water infiltration into the surrounding soils. Other burrowing squirrels, ground squirrels and marmots, similarly influence their surrounding environment.

Tree and flying squirrels likewise are important ecosystem engineers and play a significant role in the regeneration of forests around the world. Although the majority of tree and flying squirrels eat the seeds and nuts of trees, they also can act as agents of dispersal for tree seeds. Seeds and nuts that fall to the ground directly under a parent tree have very high mortality rates—first, because they compete with other seedlings and the parent tree for light and nutrients, and second, because the high abundance of seeds under a tree draws a large number of seed predators, such as bruchid beetles. Scatter-hoarding squirrels remove the seeds from under the parent tree and bury them anywhere from a few to hundreds of meters away. These buried seeds have reduced competition for nutrients and light, are less likely to be found by weevils and other predators, and are more likely to germinate in the moist, dark environment under the soil. In a good mast year, a squirrel buries an enormous number of seeds and nuts. Stephen Vander Wall estimates that "at the population or community level, the number of seeds and nuts . . . must often exceed thousands per hectare." Even though squirrels are proficient at later finding these buried seeds, they rarely, if ever, relocate every last one. Therefore, over the course of a year a good number of seeds may survive to germinate. In the spring of some years you can find a veritable carpet of seedlings where squirrels have buried nuts.

On a small scale, squirrels are important in the succession of forests after major disturbances such as fire and assist the expansion of forests into fields and previously logged areas. On a larger scale, squirrels were essential in the reestablishment of northern temperate forests after the last glacial retreat. Squirrels are considered key dispersers of at least nine genera of nut-bearing trees (*Juglans, Carya, Corylus, Fagus, Castanea, Castanopsis, Lithocarpus, Quercus, Aesculus*) and at least one species of conifer (*Pinus koraiensis*).

Many species of trees have a close symbiotic association with endorhyzal fungi. These underground fungi assist the tree to obtain and absorb nutrients from the soil, and the tree roots provide essential metabolic products to the fungi. Without these fungi, the trees may grow more slowly or not at all. Many squirrel species dig up and feed on the fruiting bodies—the truffles—of these fungi, and as a result the squirrel's feces are full of fungal spores. Thus, the squirrels disperse both the seeds and the important fungal spores. This relationship has been implicated most strongly for the fox squirrels and the long-needled pine of the coastal eastern United States, but several other species are inveterate truffle eaters and may play similar roles.

Some species of *Tamiops* and *Funambulus* in India and Southeast Asia gather nectar and in doing so become covered in pollen. By moving from one plant to another these squirrels act as important agents of pollination.

Squirrels also are popular medical research subjects. Researchers at the Institute of Arctic Biology at the University of Alaska, Fairbanks, are studying the physiology of hibernation in the Arctic ground squirrel (*Spermophilus paryii*). Blood flow to the brain, heart, and lungs of squirrels is reduced by almost 90% during deep hibernation, but hibernating squirrels seem to suffer no tissue, muscle, or nerve damage as a result. Because heart attacks and strokes in humans can result in similarly reduced blood flow, the researchers hope that by expanding their knowledge of the mechanisms underlying hibernation, they may be able to develop a better understanding of and possibly treatments for strokes, heart attacks, and neurodegenerative diseases. Woodchucks (*Marmota monax*) are susceptible to the woodchuck hepatitis virus (WHV), which is remarkably similar to the hepatitis B virus (HBV) that infects humans. Woodchucks infected with WHV exhibit a similar disease progression (liver disease, liver cancer) as seen in humans infected with HBV. Because of this, woodchucks are commonly used as animal models by researchers studying HBV. Studies using woodchucks have assisted researchers in the development of new medicines, therapies, and improved liver transplant techniques.

Squirrels: The Animal Answer Guide

Group-living, social squirrels, such as prairie dogs and most marmots, also are important research subjects. Scientists study these squirrels in hopes of learning more about the evolution of sociality, the benefits of sociality, how relatives recognize each other, social aggression, alloparenting, infanticide, communication, and many other intricacies of social living. Because these squirrels live in relatively open habitats and are active during the day, they are easy to observe.

Squirrel fur has been important for centuries, for use in clothing and as currency. During the fourteenth century, squirrel fur, in particular, from Northern Russia, became a hot commodity in Europe, and Russian towns, such as Novgorod, developed economies based almost entirely on the trade of squirrel skins. Within Novgorod, skins were used as currency or collected as tax or rent from inhabitants. Merchants would trade food, clothing, and silver with inhabitants for these skins and then sell them to clients across Europe and Great Britain. Squirrel skins were also used as money in early Finland, and the modern Finnish word for money has its roots in a word that meant squirrel skin. To this day, squirrels continue to be farm raised for their fur in parts of Europe. North American Inuit tribes still rely on the skins of Arctic ground squirrels to make coats, boots, and other items.

Chapter 9

Squirrel Problems (from a human viewpoint)

Are squirrels pests?

They can be.

In a 2003 *New York Times* article, Elisabeth Bumiller describes how squirrels living on the White House lawn killed an historic tree. It seems the squirrels stripped so much bark off the upper limbs of a buckeye tree planted by Theodore Roosevelt that it died and had to be cut down.

Bark stripping is common practice among many temperate tree squirrels and some ground squirrels. Squirrels strip away the top layer of the bark to get to the sweet, phloem tissue underneath. This occurs most commonly in early summer and can be fatal to trees, either by killing them outright or by making them susceptible to infestation by insects or fungi. Bark stripping, though relatively innocuous in wild forests, can seriously damage managed forests, which tend to have thousands of acres of a single species of similar-aged trees growing together. Squirrels prefer to debark trees of a certain diameter (less than 6 cm or 2.4 inches), and when many trees of this size are located together, the damage can be significant. Besides the risk of killing a tree, bark stripping can stunt a tree's growth and cause deformations in the wood, both of which can decrease the tree's usability by timber companies. With estimates of 30–90% damage in certain stands of trees, bark stripping can affect companies negatively. In fruit orchards, woodchucks also strip bark. In this case, they gnaw on the lower bark of trees, and studies show that trees that have been gnawed upon produce smaller and fewer fruits than ungnawed trees.

In tropical plantations tree squirrels are considered unwelcome pests, because they feed extensively on ripe fruit, cocoa, and palm nuts. In central and northern South America, the red-tailed squirrel (*Sciurus granatensis*) is a common pest on cocoa plantations, and in Southeast Asia, multiple species of *Callosciurus* commonly invade plantations. In Kenya, the striped ground squirrel (*Xerus erythropus*) frequently preys on maize seeds and crops, which can account for more than half the damage to all maize crops.

In North America, some ground squirrels also are considered agricultural pests and are known to damage many grain crops through extensive feeding. In 1910, a U.S. Department of Agriculture publication estimated that ground squirrels were responsible for more than $10 million dollars of damage per year to crops. (Today, that would be equal to approximately $200 million in damage.) A 1999 report found that Belding's ground squirrels feeding on alfalfa crops in northeastern California caused significant damage, costing farmers on average $402 per hectare.

Prairie dogs have been considered pests by ranchers in the western United States for more than a century. Ranchers claim that prairie dogs compete with livestock for food and that prairie dog burrows pose a danger to livestock and machinery. Starting at the turn of the twentieth century, ranchers, with the support of the U.S. government, began systematically killing prairie dogs using any means possible. Prairie dogs were shot, trapped, and poisoned by the hundreds of thousands. This assault, combined with the unintentional introduction of plague from Asia, resulted in an extreme reduction in prairie dog populations and their extirpation from many areas. Today, prairie dogs occupy only a fraction of their historical range (though estimates of the size of their historical range are controversial). In recent studies biologists have found that the supposed damage done by prairie dogs has been overstated, but the view of the prairie dog as pest is deep-rooted and still persists today. While two species of prairie dogs, the Mexican and Utah, have some legal protection, white-tailed and black-tailed prairie dogs do not, and attempts to gain protection for these species have been contentious and highly political.

In suburban areas, humans intentionally and unintentionally provide a plethora of tasty treats for squirrels. Buds, shoots, flowers, bulbs, fruits, and vegetables are all part of a successful suburban squirrel's diet, much to many a gardener's dismay. Squirrels readily dig up flower bulbs and will quickly remove every last tomato from a ripe vine. Squirrels frequently show up uninvited at many backyard bird feeders, quickly emptying them, and moving on. Woodchucks, also common in suburban areas, make snacks of fresh garden greens whenever possible.

Municipal parks offer havens for squirrels in the middle of urban areas. The absence of many predators, combined with supplemental feeding by park visitors, allows squirrel populations to grow relatively unchecked. A 1977 study in Lafayette Park in Washington, D.C., documented a squirrel density of approximately 12.5 squirrels per acre—one of the highest densities ever recorded! These large populations do not come without a cost, though. The hard work invested by cities planting flowers, trees, and shrubs can disappear in a few weeks. The abundant Lafayette squirrels, in one season, did $4,500 in damage by digging up 2,000 newly planted geraniums and stripping bark from trees.

Tree squirrels also have an unfortunate habit of venturing into transformers and generators, electrocuting themselves and knocking out power to surrounding communities. In October 1953, squirrel populations were so high in parts of Mississippi that roaming squirrels knocked out 20 transformers in less than a month. In 1979, a single squirrel caused $1 million in damage to an electrical substation in Pennsylvania when it wandered into a generator, triggering a fire and explosion. In 1984, an article titled "The Squirrels Who Can Zap Thanksgiving" appeared in the *Washington Post*. Reporter Tom Vesey described how eastern gray squirrels in the area were caching their nuts in electrical transformers at the top of poles, causing an increase in the number of blackouts. In the article, the electric company stated that between 1983 and 1984, there were 579 squirrel-caused electrical outages in that region.

In his book, Fred Barkalow describes the squirrel as an "energetic nibbler," and this is an apt depiction. Nothing appears immune from the investigations of the squirrel's teeth—electrical wires, telephone cables, house siding, gutters, roof shingles, cardboard boxes, lawn chairs, bird feeders, shutters, insulation. You name it, a squirrel will nibble it. This habit can cause significant damage, and homeowners around the world have cursed the bushy-tailed perpetrators. Flying squirrels commonly find their way into attics or behind walls and set up nests, where they nibble through insulation and wire coverings, leaving bare wires that can increase the risk of fire.

How do I keep squirrels away from my . . . ?

If you are truly committed to keeping squirrels away from your homestead, then the best advice is to not put *any* food out for birds, pets, or other animals. Keep garbage in spill-proof containers and thoroughly clean the outside of barbecue grills. Do not plant vegetables, flowers, or other vegetation that may produce anything squirrels will find tasty. Although these steps may not completely eliminate squirrels from your property, they will probably reduce the number of squirrels.

If these steps seem a bit severe to you, there are many other ways to limit the effect of squirrels on your bird feeder, your garden, and your home.

Northern palm squirrel (*Funambulus pennantii*), South Asia. Photo © S. N. Naik

Layard's squirrel (*Funambulus layardii*), South Asia. Photo © Rajith Dissanayake

South African ground squirrel (*Xerus inauris*), Africa. Photo © Jane Waterman

Harris's antelope squirrel (*Ammospermophilus harrisii*), North America and Mexico. Photo © Robert Shantz, www.rshantz.com

Round-tailed ground squirrel (*Spermophilus tereticaudus*), North America and Mexico. Photo © Jim Hughes, www.scarysquirrel.org

Thirteen-lined ground squirrel (*Spermophilus tridecemlineatus*), North America. Photo © Phil Myers

Columbian ground squirrel (*Spermophilus columbianus*), North America. Photo © Phil Myers

Belding's ground squirrel (*Spermophilus beldingi*), North America.

Photo © Phil Myers

Golden-mantled ground squirrel (*Spermophilus lateralis*), North America. Photo © Gregg Elovich, www.scarysquirrel.org

Gray-collared chipmunk (*Tamias cinereicollis*), North America. Photo © Robert Shantz, www.rshantz.com

Least chipmunk (*Tamias minimus*), North America. Photo © Phil Myers

(Left) Lodgepole chipmunk (*Tamius speciosus*), North America. Photo © Douglas Herr **(Right) Woodchuck (*Marmota monax*), North America.** Photo © Joe Kosack, Pennsylvania State Game Commission

Yellow-bellied marmot (*Marmota flaviventris*), North America. Photo © Gregg Elovich, www.scarysquirrel.org

(Left) Hoary marmot (*Marmota caligata*), North America. Photo © U.S. Fish and Wildlife Service (Right) Vancouver Island marmot (*Marmota vancouverensis*), North America. Photo © Andrew A. Bryant

Olympic marmot (*Marmota olympus*), North America. Photo © Andrew A. Bryant

White-tailed prairie dog (*Cynomys leucurus*), North America. Photo © Robert Shantz, www.rshantz.com

Black-tailed prairie dog (*Cynomys ludovicianus*), North America and Mexico. Photo © Gregg Elovich, www.scarysquirrel.org

BIRD FEEDERS. There are hundreds of bird feeders on the market that claim to be "squirrel proof." And, in all likelihood, there are hundreds, if not thousands, of people who have bought these bird feeders only to find a crafty squirrel blissfully munching away on it one morning. Although there may always be that "one" squirrel who can outwit any squirrel-proof contraption, there are strategies you can use to reduce the chances that a squirrel will gain access to a bird feeder.

Placement of the feeder is key. Squirrels are amazingly adept acrobats. Their hind feet can rotate almost 180 degrees, allowing them to hang off of the most precarious perches to grab food with their fore feet. They also can jump significant distances, especially if there is a tasty reward on the other side. The trick is to eliminate places from which the squirrel can hang, jump, or stretch to reach the feeder. This means that the feeder should be at least 6 feet off the ground and at least 8 feet horizontally (preferably farther) from the nearest launching point (e.g., tree branch, picnic table, roof, dog house).

Even an isolated feeder is not fully protected. Squirrels can shimmy up poles and tiptoe across even very thin wire. Stringing barriers along wire hung feeders is one commonly used trick. Bill Adler Jr., in his book *Outwitting Squirrels*, suggests hanging a feeder on a horizontal wire. Then, on either side of the feeder, string several record albums, which should be high and wide enough to prevent the squirrel from climbing over. Large aluminum pie plates probably would work just as well. We have used plastic tubing, cut into half-inch lengths, which were then threaded onto the wire. Each piece of tubing rotated freely and easily on the wire, which made it very difficult for squirrels to walk across, until an ice storm cemented them all together.

Baffles are useful additions to pole-mounted bird feeders and come in a variety of shapes—half-spheres, cones, pointed, slanted, among others. They all have the aim of blocking and preventing the squirrel access to the bird seed. Baffles can be good protection, but they are not foolproof as hungry squirrels are not easily deterred and can wiggle, squirm, squeeze, and contort their way around a host of obstacles. Greasing poles is not recommended, because substances such as petroleum jelly can adversely affect the insulating ability of bird feathers and animal fur.

Another option is to fill your bird feeders with foods that squirrels do not like—thistle and safflower seeds are two options. This might limit the types of birds you attract to your feeders, however. Anecdotal evidence suggests that mixing cayenne pepper in with the bird seed will keep squirrels away, because squirrels can taste it but birds cannot.

Finally, you can distract squirrels with a feeding station of their own. Set up an easily accessed feeding station, located a good distance from any bird feeder, and fill it with foods irresistible to a hungry squirrel—sunflower

An eastern fox squirrel raids a bird-feeder in the rain. Keeping squirrels out of their birdfeeders has consumed many a backyard bird lover.

Photo © Phil Myers

seeds, unsalted peanuts, and acorns. Squirrels typically go for easy to obtain food, and so a supplemental feeding station in combination with the preventative measures above, should limit extraneous raids of your bird feeders.

GARDENS, FLOWERBEDS, AND POTTED PLANTS. In Lafayette Park, across from the White House, squirrels did $4,500 worth of damage one spring by digging up and eating most of the newly planted geraniums. On a much smaller scale, many backyard gardens and flowerbeds suffer similar abuses every spring and summer. Keeping squirrels out of your gardens and flowerbeds is much like keeping them out of your bird feeder—an ongoing battle. You may never be entirely successful, but you can take preventative steps to decrease the damage. One of the most common techniques is to cover newly planted bulbs and seeds with wire mesh, which will prevent squirrels (or other animals) from digging them up. Once the seeds have sprouted, you can cover the entire plant with some sort of netting or cage to prevent unwanted access.

Another trick is to spray plants with a cayenne pepper and water solution, but this wears off quickly and must be reapplied often, and using it on vegetables you plan to eat is not recommended. You can also buy commercial sprays that claim to deter squirrels, but the efficacy of these products is undetermined. Some other tricks that have been claimed to work with varied success include spreading blood meal on your flowerbeds, placing dog hair around your plants, and even using mothballs.

Ground squirrels, particularly woodchucks, can be very difficult to keep out of a vegetable garden, because they easily dig under a fence. If you take flashing or some other hard barrier and sink it into the ground a foot or so beneath a fence, you may be able to defeat them.

Most species of squirrels are active during the day, so they are easily spot-

Squirrels: The Animal Answer Guide

ted on their raids and therefore most commonly blamed for damage. There are many nocturnal animals that also may be eating your plants, including raccoons, possums, mice, rats, rabbits, skunks, bears, foxes, and deer.

ATTICS, WALLS, AND PORCHES. Squirrels cause damage to homes in two main ways: by gnawing on siding, roof tiles, or wiring or by nesting inside an attic or crawl space. The gnawing habit of squirrels is hard to prevent, as squirrels have ever-growing incisors and therefore must gnaw on hard objects (usually nuts) to keep them well worn and sharp. It just so happens that siding, roof tiles, porch railings, and wooden steps all provide hard surfaces ideal for chewing.

The squirrel gnawing habit becomes most worrisome if they strip the insulation off of electrical wires.

The warm, dry protection offered by an attic will attract female squirrels looking for safe nesting sites. In North America, flying squirrels are the most common attic inhabitants, but eastern gray and eastern fox squirrels also will take advantage of them from time to time. Squirrels commonly gain entrance to an attic or crawl space through openings between the roof and the walls, through gutter spouts, and through ventilation fans. Squirrels also can gain entrance to houses through chimneys, but once inside the chimney they rarely can get out. The best way to keep squirrels out of a house is to block all of their entrances. You should walk around the house and take stock of any opening they may be able to use. Even openings that seem too small should be noted, because squirrels can get through very small openings. Any holes or openings should be patched, making sure that (1) the squirrels are no longer in the house and (2) that no babies are left inside the house. Also, all chimneys and vents should be capped.

Finally, another complaint is about squirrels chewing on pillows and rugs left out on porches. Most likely, these squirrels are gathering soft material to use in lining their nests. Leaving out some old pillow batting or loose wool for the squirrel to use instead may help to keep cushions intact.

TRAPPING SQUIRRELS. In the United States, wildlife is the property of the state, and many states consider that squirrels are game animals to be managed. They therefore have strict regulations on hunting and trapping, which may or may not pertain to your situation. It is wise to check the regulations before trapping or killing squirrels.

Live trapping and relocating nuisance squirrels, though common, is not a good idea. Although relocating squirrels to a "nice home in the country" seems humane, relocated squirrels rarely, if ever, survive successfully in new territory. A three-year study by squirrel expert Van Flyger and colleagues found that 97% of suburban squirrels that were relocated to a park either died or "disappeared" within the year. Squirrels that are relocated to new, unfamiliar areas are at high risk for predation because they do not have established nests and hiding places. Depending on the season, they are at high risk for starvation, because they do not have established food stores to exploit. Finally, they are also at high risk of aggressive encounters with other squirrels. Relocation is exceedingly stressful to the animal and, as the study above has shown, rarely successful. In addition, when you remove squirrels from your neighborhood, new squirrels soon enter the area to fill the recent vacancy. Laws in many states and counties prohibit the trapping, transportation, and release of live animals without a license.

Finally, and this cannot be said enough, NEVER USE POISON! It is almost impossible to ensure that poison will be ingested only by the intended pest animal. In most cases other wildlife will ingest the poison, and there is a very high risk that your neighborhood or family pets will ingest the poison as well. Also at high risk are the predators that feed on the animals that ingest the poisoned bait. Eagles, hawks, falcons, owls, snakes, carrion feeders (i.e., buzzards, crows), foxes, and mountain lions are just a few of the many predators that inadvertently die each year because they ingested poisoned animals.

Are squirrels vectors of human disease?

Squirrels are not considered important vectors (or carriers) of any human diseases. Rabies is a case in point. Squirrels can get rabies, but it is not very commonly documented. According to the Centers for Disease Control and Prevention (CDC) squirrels are almost never found to be infected and no case of squirrel to human transmission of rabies has been documented in the United States. An upsurge in reports occurred between 1980 and 1984 in the mid-Atlantic states, where there were more than 50 cases of rabies reported in woodchucks (*Marmota monax*), in association with an epizootic among raccoons (*Procyon lotor*). During the next decade there were almost 300 additional cases countrywide, mostly in the East. In Maryland 15 cases of human exposure to rabies due to woodchucks were reported by the Health Department, 1981–1986. To put this in perspective, we note that woodchucks accounted for less then 1% of the animals found rabid in Maryland during this period.

Although some other diseases squirrels carry can be transmitted to humans, the transmission does not commonly occur directly from squirrel to human. Two examples of this are the West Nile virus and Lyme disease. In both cases, even though the squirrel carries the pathogen, humans can only get infected through the bite of the actual vector—the mosquito and tick, respectively.

Chapter 10

Human Problems (from a squirrel's viewpoint)

Are squirrels endangered?

Yes, some species of squirrels are definitely endangered, while other species definitely are not. Habitat loss is perhaps the most important factor contributing to the decline of squirrel populations around the world. Forests and grasslands are disappearing to make way for farming, grazing, and human development. Although some species of squirrels have proven adept at living alongside humans, many squirrels will not be able to survive in this altered environment. The following example demonstrates that the situation may be exceedingly complex.

Vancouver Island marmots are one of the most endangered mammals on the planet. Endemic to Vancouver Island in British Columbia, there are only 35 still living in the wild and a further 93 marmots in captivity. These marmots live at elevations of 1,000 meters (3,281 feet) or higher and depend on treeless meadows and good soil for burrowing. The causes for the decline of the Vancouver Island marmot are multiple and include human disturbance, weather, predation, disease, and hunting. In the 1960s and 1970s areas of Vancouver Island were logged extensively. These logging activities cleared land and created artificial meadows that seemed to benefit the Vancouver Island marmot. Since marmot numbers peaked at 350 animals in the 1980s, however, they have declined. Studies have shown that marmots living in these clear-cut habitats produce fewer dispersing young than those marmots that live in natural meadows. Because fewer young are dispersing, the numbers of marmots living in a small area within and around these habitats are larger, thus increasing the risk for severe weather or predation to wipe out a large number of marmots at one time. In addi-

Table 10.1. Endangered and threatened squirrel species

Red List—International Union for Conservation of Nature and Natural Resources
(IUCN)

- Critically endangered

  Namdapha flying squirrel (*Biswamoyopterus biswasi*), Southeast Asia

  Sumatran flying squirrel (*Hylopetes winstoni*), Southeast Asia

  Northern Idaho ground squirrel (*Spermophilus brunneus brunneus*), North America

  New Mexico least chipmunk (*Tamias minimus atristriatus*), North America

  Hidden forest chipmunk (*Tamias umbrinus sedulus*), North America

  Mount Graham red squirrel (*Tamiasciurus hudsonicus grahamensis*), North America

  Vincent's bush squirrel (*Paraxerus vincenti*), Africa

- Endangered

  Nelson's antelope ground squirrel (*Ammosphermophilus nelsoni*), North America

  Mexican prairie dog (*Cynomys mexicanus*), North America

  Woolly flying squirrel (*Eupetaurus cinereus*), South Asia

  Particolored flying squirrel (*Hylopetes alboniger*), South Asia

  Sipora flying squirrel (*Hylopetes sipora*), Southeast Asia

  Vancouver Island marmot (*Marmota vancouverensis*), North America

  Northern Palawan tree squirrel (*Sundasciurus juvencus*), Southeast Asia

  Complex toothed flying squirrel (*Trogopterus xanthipes*), North Asia

  Prince of Wales flying squirrel (*Glaucomys sabrinus griseifrons*), North America

  Idaho ground squirrel (*Spermophilus brunneus*), North America

- Vulnerable

  Irrawady squirrel (*Callosciurus pygerythrus*), Southeast Asia

  Anderson's squirrel (*Callosciurus quinquestriatus*), Southeast Asia

  Carruther's mountain squirrel (*Funisciurus carruthersi*), Africa

  Carolina northern flying squirrel (*Glaucomys sabrinus coloratus*), North America

  Virginia northern flying squirrel (*Glaucomys sabrinus fuscus*), North America

  Montane long-nosed squirrel (*Hyosciurus heinrichi*), Southeast Asia

  Lowland long-nosed squirrel (*Hyosciurus ileile*), Southeast Asia

  Mentawi flying squirrel (*Iomys sipora*), Southeast Asia

  Four-striped ground squirrel (*Lariscus hosei*), Southeast Asia

  Menzbier's marmot (*Marmota menzbieri*), North Asia

  African pygmy squirrel (*Myosciurus pumilio*), Africa

  Cooper's mountain squirrel (*Paraxerus cooperi*), Africa

  Red bush squirrel (*Paraxerus palliatus*), Africa

  Swynnerton's bush squirrel (*Paraxerus vexillarius*), Africa

  Small Travancore flying squirrel (*Petinomys fuscocapillus fuscocapillus*), South Asia

  Secretive dwarf squirrel (*Prosciurillus abstrusus*), Southeast Asia

  Indian giant squirrel (*Ratufa indica*), South Asia

  Grizzled giant squirrel (*Ratufa macroura*), South Asia

  Forrest's rock squirrel (*Sciurotamias forresti*), South Asia

  Southern Idaho ground squirrel (*Spermophilus brunneus endemicus*), North America

European ground squirrel (*Spermophilus citellus*), Europe
Franklin's ground squirrel (*Spermophilus franklinii*), North America
Mohave ground squirrel (*Spermophilus mohavensis*), North America
Speckled ground squirrel (*Spermophilus suslicus*), Europe
Washington ground squirrel (*Spermophilus washingtoni*), North America
Jentink's squirrel (*Sundasciurus jentinki*), Southeast Asia
Palawan montane squirrel (*Sundasciurus rabori*), Southeast Asia
Samar squirrel (*Sundasciurus samarensis*), Southeast Asia
Selkirk least chipmunk (*Tamias minimus selkirki*), North America
Palmer's chipmunk (*Tamias palmeri*), North America
Organ Mountains chipmunk (*Tamias quadrivittatus australis*), North America

United States Fish and Wildlife Service Endangered Species Listings

- Endangered
  Carolina northern flying squirrel (*Glaucomys sabrinus coloratus*)
  Virginia northern flying squirrel (*Glaucomys sabrinus fuscus*)
  Delmarva Peninsula fox squirrel (*Sciurus niger cinerus*)
  Mount Graham red squirrel (*Tamiasciurus hudsonicus grahamensis*)

- Threatened
  Northern Idaho ground squirrel (*Spermophilus brunneus brunneus*)
  Utah prairie dog (*Cynomys parvidens*)

- Candidates for listing
  Palm Springs round-tailed ground squirrel (*Spermophilus tereticaudus chlorus*)
  Southern Idaho ground squirrel (*Spermophilus brunneus endemicus*)
  Washington ground squirrel (*Spermophilus washingtoni*)

tion logging roads increase the ease with which predators can travel from place to place.

The Canadian government has taken action to protect the Vancouver Island marmot, first listing it as threatened in 1979. A recovery team was established in 1988, and an official recovery program was started in 1993. The recovery of the Vancouver Island marmot is a multilayered process that includes captive breeding programs, reintroductions, population monitoring, ongoing research, habitat protection and manipulation, contingency plans (in case of environmental catastrophe), fundraising, and public communication. The Canadian government even named May 1, "Marmot Day," with the idea that the endangered marmot is yelling, "Mayday! Mayday!" or to French Canadians, "M'aidez! M'aidez!" At this time, recovery efforts appear to be successful. Andrew Bryant, a senior researcher for the Marmot Recovery Program, stated that in 2005 the total number of Vancouver Island marmots has reached 120 marmots for the first time since 1998.

A female Vancouver Island marmot (*Marmota vancouverensis*) named Oprah interacts with her young at Haley Lake Ecological Preserve on Vancouver Island. She was first captured and marked in 1987 as a 2-year-old. When this photo was taken, she was 8 years old and had just had her third litter of pups. Her pup in this picture is probably 5–6 weeks old. Oprah lived to age 9, which is the maximum age so far recorded for a Vancouver Island marmot wild female. Photo © Andrew A. Bryant

Another big obstacle facing the conservation of squirrel populations is the enormous lack of data on many species of squirrels, in particular, squirrels living in remote areas, such as the Himalayas or in the forests of Southeast Asia. Regrettably, habitat destruction is outpacing data collection in many of these areas, and is severe in parts of Southeast Asia. One of our colleagues, Illar Muul, spent many years in Southeast Asia, where he collected squirrels for epidemiological studies. He found a rare species of pygmy flying squirrel, *Petaurillus kinlochii*, in one site in Selangor, the only state in Malaysia where this squirrel has been found. Upon returning to the site in 2004, he discovered the forest had been cut down. Whether or not *Petaurillus kinlochii* still exists anywhere is not known. Similarly, in many other areas, squirrels may be threatened but we do not have enough information to know.

Two other very significant factors affecting squirrel populations are climate change and invasive species. We treat these next as separate questions.

Will squirrels be affected by global warming?

Yes, squirrels will be affected by climate change, as will almost all other animals and plants. Change has been a consistent theme throughout the earth's history. Over the course of millions of years, land masses have come together and split apart, mountains have formed and been eroded away, seas have been created and then disappeared, and glaciers have advanced and retreated. As the geography changed so too did the earth's climate, and these variations in geography and climate have fueled the evolution and extinction of species.

Global warming is an increase in earth's average temperature, which causes many other changes in the earth's climate. We know that throughout

Squirrels: The Animal Answer Guide

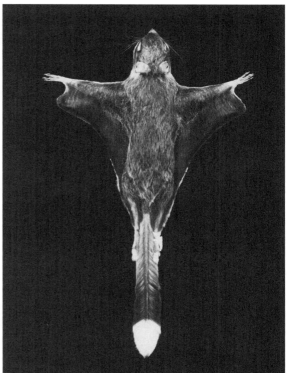

(Left) A female Vancouver Island marmot at Haley Lake Ecological Preserve on Vancouver Island. This female was 4 years old when the photo was taken and lived for at least two more years, after which she disappeared. Her colony became extinct in 1998, but in 2004 captive-bred marmots were introduced to the site. Photo © Andrew A. Bryant
(Right) A museum specimen of the Selangor pygmy flying squirrel (*Petaurillus kinlochii*) from Southeast Asia.

the earth's history the planet's average temperature has increased at various times and decreased at others. The current trend in global warming is extraordinary, however, because it is happening as the direct result of human actions—namely, our increased emissions of gases such as carbon dioxide—and it is occurring much more rapidly than global warming in the past. Scientists do not know what the end effect of global warming will be; yet biologists conducting long-term field studies are already noticing behavioral and other changes in species that seem to result from global warming.

David Inouye, of the University of Maryland, has been doing research at the Rocky Mountain Biological Laboratory in Crested Butte, Colorado, for 33 years. When Inouye and his colleagues looked at some of the long-term data, they saw that yellow-bellied marmots (*Marmota flaviventris*) at the site are emerging from hibernation 38 days earlier than they were 23 years ago. They hypothesize that this earlier emergence is due to warmer spring air temperatures.

Denis Reale, of McGill University, Stan Boutin, of the University of

Alberta, and their colleagues noticed similar behavioral changes in a North American red squirrel (*Tamiasciurus hudsonicus*) population in the southwest Yukon. Average spring temperatures in that area have increased 2°C (4°F) over the past 27 years, and in response to this the local population is now breeding 18 days earlier than 10 years ago. The warmer temperatures have also increased the spruce cone crops, which North American red squirrels feed on, increasing food abundance over the animal's lifetime.

Both of these populations of squirrels are robust, and it seems that they are adapting to the changing temperatures. It is unclear what effect global warming may have on endangered and threatened populations of squirrels. One example is the Mount Graham red squirrel (*Tamiasciurus hudsonicus grahamensis*), which is restricted to elevations above 3,200 m in the Pinaleño Mountains in Arizona. Between 1986 and 2001, the population of these squirrels never exceeded 580 individuals. During the last glacial period the Mount Graham red squirrel had a much larger range because of the pervasive cool, moist environment and was connected to more northern populations. As the glaciers retreated and the climate warmed and dried, the squirrels retreated up the mountains to find suitable habitat and since have been isolated from other populations of squirrels for about 10,000 years. Continued global warming threatens to eliminate what is left of the cool, moist environment that Mount Graham red squirrels depend on for survival. Many Alpine populations of marmots also face the same risk as temperatures warm and their habitat changes.

Another risk of global warming is disease. New parasites may flourish in a warmer environment, affecting squirrels that have no resistance to them. One potential example of this was examined by Peter Weigl and colleagues at Wake Forest University. He studied the nematode parasite *Strongyloides robustus* and its effects on populations of northern and southern flying squirrels that co-occur in areas of the Appalachians in North America. Although southern flying squirrels readily carry and transmit *Strongyloides robustus*, northern flying squirrels are a naïve host—rarely becoming infected. When infected, however, northern flying squirrels often die, and southern flying squirrels rarely suffer any harmful effects. Currently, where northern flying squirrels and *Strongyloides* co-occur temperatures are unfavorable to the parasite, as it requires more consistent warm temperatures to successfully reproduce in large numbers. This keeps infection rates very low among the northern flying squirrels. If temperatures increase, *Strongyloides* probably will become more successful and infections among northern flying squirrels will increase. Because northern flying squirrels do not have defenses against the parasite, death rates will probably be high. The result may be a die-off of the northern flying squirrels in this range.

Squirrels: The Animal Answer Guide

Are squirrels ever invasive species?

An introduced species is a species not native to the ecosystem, region, or country in which it now resides. Species are introduced either intentionally or unintentionally. These species are sometimes beneficial, sometimes benign, and sometimes have a long-lasting, negative effect on the ecology of their new "home." These latter, highly successful and prolific introduced species are commonly referred to as invasive species.

The most common introduced squirrel species is the eastern gray squirrel (*Sciurus carolinensis*). This squirrel species has been introduced to many places across the globe either for novelty or sport, and in some areas it is clearly an invasive species. Perhaps the best documented is its introduction to the United Kingdom from North America in the mid-1880s. The eastern gray squirrel quickly became established across most of the United Kingdom and, in conjunction with deforestation, has been identified as the primary cause of the demise of the native Eurasian red squirrel (*Sciurus vulgaris*), a smaller species found in the United Kingdom and across most of Eurasia. The reasons why the Eurasian red squirrel is being replaced by the eastern gray squirrel are not completely understood. Luc Wauters, of the University of Insburia, discovered that in areas where Eurasian red and eastern gray squirrels coexist, the Eurasian red squirrel has a lower daily energy intake than in areas where only Eurasian red squirrels exist. This may mean that the eastern gray squirrel is out-competing the Eurasian red squirrel for high-quality food or that the eastern gray squirrel is stealing food from Eurasian red squirrel caches. Disease spread by the eastern gray squirrel also seems to be having a significant impact on the decline of the Eurasian red squirrel, at least in the United Kingdom. Daniel Tompkins, at the University of Stirling, and colleagues have looked at one likely culprit, the parapox virus. This virus, carried by the eastern gray squirrel, spread to the Eurasian red squirrel as the eastern gray squirrel's range expanded across the United Kingdom. While the eastern gray squirrel is immune to the virus, the Eurasian red squirrel has limited or no immunity to it. The virus, which causes ulcers, skin lesions, and often death, has killed numerous Eurasian red squirrels across the country.

Eastern gray squirrels also have been introduced into other areas. They were introduced into Italy three different times: in 1948 to Piedmont, in 1966 to Genoa, and in 1994 to Trecate. The Trecate population no longer exists, the Genoa population is confined to a relatively small area, but the Piedmont population has expanded to an area of 880 km² (339.8 square miles). A large-scale plan to eradicate the eastern gray squirrel from Italy was stalled in 1997, and it is feared that the eastern gray squirrel will continue to expand into mainland Europe and the rest of Eurasia, causing devastation to native Eurasian red squirrel populations similar to what has occurred in the United Kingdom.

In 1900 the eastern gray squirrel was brought to Groote Schuur in South Africa, and a population still exists, though it is limited to the western Cape. Eastern gray squirrels also have been introduced to parts of western Canada and the northwestern United States. At one time, a population of introduced eastern gray squirrels existed in Melbourne, Australia, but it has since gone extinct.

Because of its success establishing itself in new environments and the risk it poses to native squirrel species, the eastern gray squirrel has been included on the list of "100 of the World's Worst Invasive Alien Species" put out by the Invasive Species Specialist Group of the World Conservation Union (www.iucn.org).

Several other squirrel species also have been introduced to new areas, though on much smaller and (so far) less destructive scales than the eastern gray squirrel, so they are not yet considered invasive species. Fox squirrels (*Sciurus niger*) have been introduced into California, Oregon, Washington, Idaho, and Montana. Finlayson's squirrels (*Callosciurus finlaysonii*), native to Southeast Asia, were introduced to northwest Italy in 1988, and now about 50 individuals are established there. Siberian chipmunks (*Tamias sibiricus*), native to Scandinavia, western Russia, and Asia, exist in isolated populations in France, Italy, Germany, Netherlands, and Austria—most established by escaped pets. A population of Indian palm squirrels (*Funambulus pennanti*) exists in Perth, Australia, and the Southeast Asian red-bellied tree squirrel (*Callosciurus erythraeus*) has been introduced to Japan, France, and Argentina.

Native animals are occasionally reintroduced to areas where they have gone extinct. For example, historical climate change severely shrank the range of the Alpine marmot (*Marmota marmota*) in Europe. Starting in 1860, reintroduction programs for the Alpine marmot began in the Austrian Alps and in 1948 they began in the French Pyrenees. Thanks to these reintroductions, stable populations of Alpine marmots have successfully recolonized significant areas of their previous range.

Do people hunt and eat squirrels?

Yes, squirrel hunting is a common practice with a long history. It is hard to imagine any type of hunting more linked with American history and tradition than squirrel hunting. The famous Kentucky long rifle, developed in the early 1700s by German gunsmiths in Pennsylvania, was often referred to by many early pioneers as their "squirrel rifle." Superbly handmade and extremely functional, the rifles were amazingly accurate in the hands of these early hunters.

Squirrels: The Animal Answer Guide

Evidence of exactly how capable these master riflemen were was seen later that century when Pennsylvania placed a bounty of three pence each on the common bushytails. Many men simply quit work and went squirrel hunting, and within one year the Pennsylvania treasury was drained of eight thousand pounds sterling.

The guns have changed, but the hunting of squirrels is still common and many states have squirrel seasons and bag limits for hunters. Squirrels are still one of the most popular game animals among hunters in the eastern United States. The United States Fish and Wildlife Association reported that there were 2,119,000 squirrel hunters in 2001 and estimated that the average hunter spends at least $2,000 per season.

Regarding eating squirrels, a search on the Internet for "squirrel recipes" yielded 166,000 hits. Native Americans undoubtedly hunted squirrels and ate them before the colonists arrived. The Hopi and Navajo tribes have recipes for baked prairie dogs. They recommend animals taken in the spring, presumably because prairie dogs put on fat and become excessively greasy later in the year. Because marmots are large, they have also been hunted for food, and they continue to be popular among persons who eat wild game. In Mongolia, tarvaga (*Marmota sibirica*) hunts are conducted on horseback and are popular pastimes.

In many other countries, hunting of squirrels is determined by the price of shotgun shells relative to the value of bush meat, which usually means that large squirrels are hunted and small species are ignored by hunters, although small squirrels are sometimes hunted by children. In Africa, monkeypox virus occurs most frequently among 5- to 9-year-old children in small villages, because they hunt and eat squirrels.

Are squirrel-hair brushes actually made of squirrel hair?

Yes, they are made from the hairs of the tails of squirrels and the brushes are used in painting and in the application of cosmetics. We are familiar with three different species that are used: the North American red squirrel of Canada (*Tamiasciurus hudsonicus*), the Eurasian red squirrel (*Sciurus vulgaris*), and the northern palm squirrel (*Funambulus pennanti*).

In Rajasthan, India, the northern palm squirrel is captured in the spring of the year, the long lateral hairs on the tail are cut off, and these are used to make brushes for artists. The squirrel is released. The hairs are carefully selected and a small bundle of them are slipped into a thin tube, the ferrule, adjusted so that those hairs protruding form a suitable brush, glued in place, and a handle is attached to the other end of the ferrule to complete the brush.

Most squirrel hair used in art or cosmetic brushes comes from Canadian populations of North American red squirrels and Russian populations of Eurasian red squirrels. Three kinds of hair are recognized among the Russian squirrel-hair brushes. These seem to originate from three different areas in Russia, where the squirrels have different colored hairs. The Talahoutky hairs are brown, Kazan hairs are more brown-black, and Sacamena hairs are blue-black. In contrast, the Canadian squirrel hairs are yellow-brown, with dark tips. Surprisingly, even among the Russian squirrels there are reputed differences in quality of the hairs, and the Sacamena hairs are considered the best for watercolor brushes. The differences seem to relate to the thickness of the hairs and their springiness, which affect how they hold paint and how it flows to the tip of the hairs.

Why do so many squirrels get hit by cars?

There are at least two answers to this question. First, tree squirrels, when chased by predators, use a strategy of dodging to and fro in an attempt to out-maneuver the predator. When they cross a street and react similarly to the threat of an automobile, a successful strategy quickly becomes a fatal strategy as they dodge directly under a wheel. It is instructive to compare this strategy with the straight-line, mad-dash strategy of cats, which is a far more successful way to avoid automobiles. Young tree squirrels seem particularly vulnerable to automobiles. With age, the squirrels find ways to avoid automobile predation by crossing roads on wires or by other means. Curt Sabrosky, an entomologist and former colleague at the Smithsonian, described the road-crossing technique of a gray squirrel just south of the White House. The squirrel waited on the grass near a crosswalk until people began to cross the street, and then it crossed the street behind them. We must confess that we don't know if this was an accidental observation or the usual strategy of the squirrel.

In other cases, squirrels are commonly hit by cars because the distribution of resources causes the squirrels to cross roads frequently. The lush median strip covered with grass and clover is a temptation that has lured many a woodchuck to its death on the highway. The presence of a nut-bearing tree on the opposite side of the street from a chipmunk's burrow will cause the animal to make numerous trips back and forth across the street. It is especially a death trap if nuts fall on the street and the shells are broken open by the tires of automobiles, because the chipmunk will stop to collect all the pieces of the nut meat, right in the path of the next automobile.

Chapter 11

Squirrels in Stories and Literature

What roles do squirrels play in religion and mythology?

Squirrels are incorporated into religious stories and myths, usually for two reasons. One is to promote a moral, and the other is to explain something, usually about the squirrel itself. In the Indian epic, the Ramayana, a squirrel is depicted as assisting in the construction of a bridge from India to Sri Lanka, in preparation for the invasion to free Sita, Rama's wife, from Ravana, who had captured and imprisoned her. In various versions of the story, the contributions of the squirrel are berated by the monkeys, or the contribution proves to be invaluable. In any case, the squirrel is rewarded by Lord Rama, who strokes the squirrel on the back with three fingers, resulting in the three stripes on the Indian palm squirrel (*Funambulus palmarum*). The story is recognized in India for its moral, that even the smallest contributions should be valued, and it's regularly cited in appropriate circumstances.

The theme, "how the squirrel got its stripes," is found in American Indian legends, some with a moral, some without. A common story is that the chipmunk teased a bear, was chased by it and barely escaped, with the bear's claws raking its back in such a way as to leave the stripes. Among the Iroquois, the story involved the arrogance of the bear and its pride in its strength. The chipmunk asked if the bear could stop the sun from rising, and when the bear failed to do so, the chipmunk ridiculed it and just barely escaped its irate pursuer. The chipmunks still have the stripes to remind them not to tease other animals. In a Seneca myth, a grandmother and her granddaughter shared a bearskin blanket, which came alive and chased them. They escaped with scratches, and since they were chipmunks, chipmunks still retain the stripes in memory of the event.

In a completely contrary role, the squirrel Ratatosk in Nordic legend, is a messenger conveying malicious gossip and insults between the terrible dragon of the underworld, Nidhogg, at the root of the world ash tree, Yggdrasil, and the eagle at the very top of the tree. Ratatosk is portrayed as traveling easily between all the worlds of Nordic mythology. Despite the unpleasant reputation of Ratatosk, the name is frequently used today in Scandinavia for gossip columns and the like, for a shamanistic group in Stockholm, a computer program, and even a cattery in Cape Town, South Africa. The role of Ratatosk is probably derived from the habit of the European tree squirrel (*Sciurus vulgaris*) to give a scolding alarm call in response to danger. It takes only a little imagination for you to think that the squirrel is saying nasty things about you.

It is perhaps a residual influence of the story of Ratatosk that caused the practice of burning squirrels at the Easter bonfire in medieval times in Europe, as reported by Sir James Frazer in *The Golden Bough*. Frazer gives no explanation for this practice, so we can only speculate on whether squirrels were considered evil in the era of tree worship, because they feed on seeds and nuts, or whether the practice resulted from a desire to burn a malicious gossip. J. R. R. Tolkien seems to have incorporated evil, or at least unsavory, squirrels into his story *The Hobbit*, which describes the squirrels of Mirkwood as black and bad tasting. One may wonder if this idea derived from *The Golden Bough*, because Tolkien was a medieval scholar who had read Frazer's monograph.

Tolkien also was very familiar with the Finnish epic, the Kalevala. In it, there are passing references to squirrels, but they do not do anything significant and nothing evil or malicious. It is obvious that the authors were well acquainted with the behavior of the European red squirrel, and there is an interesting allusion to a white squirrel born to a virgin. They probably had firsthand knowledge of albinism in this species.

In a Cherokee story, the origin of bats and flying squirrels was explained. Two small mouselike creatures wanted to play in a game of ball between the birds and the mammals. The larger mammals laughed at them, but the birds took pity on them. They took some old woodchuck skin from a drum and fashioned a pair of wings for one of them, which was the bat. For the second one, there was only enough skin left to stretch between the fore and hind limbs, that is why the flying squirrel can glide, but not truly fly. In the subsequent game between the birds and mammals, the birds won with the special assistance of their two new teammates.

We conclude with a more modern myth generated in nineteenth-century America, Groundhog Day, which is based on a groundhog emerging from hibernation on the second of February and looking for its shadow. Legend says that if it sees its shadow, the groundhog returns to hibernation, because there will be six weeks more of winter. If it does not see its shadow, then it will stay out of hibernation, because winter is ending. The tradition reputedly originated in Europe from the belief that the extent of winter weather remaining could be predicted by whether Candlemas Day was cloudy or clear. According to an old English song:

> If Candlemas be fair and bright,
> Come, Winter, have another flight;
> If Candlemas brings clouds and rain,
> Go Winter, and come not again.

In Germany, the arbiter of this decision was the hedgehog. If it saw its shadow upon emerging from hibernation, more harsh winter weather was to be expected. If it did not, spring would soon come. German migrants to Pennsylvania, the Pennsylvania Dutch, had no hedgehogs to decide the matter, but groundhogs (also called woodchucks) were plentiful. In 1886, the tradition was started in Punxsutawney, Pennsylvania, by routing out a groundhog from hibernation, to see if it would observe its shadow. It is now a favorite annual festival in Punxsutawney, presided over by "Punxsutawney Phil," the groundhog that makes the decision. Since 1886, Phil has seen his shadow 87% of the time and is reported to have been 100% accurate in his weather predictions. We do not know what that says about Pennsylvania winters.

What do squirrels have to do with the Cinderella story?

In the story of Cinderella, her fairy godmother provided her with a coach and horses, fancy dress, and all the other accoutrements of a princess, including glass slippers. If you think about how uncomfortable glass slippers would be, you will realize that this was a very strange choice of footwear. As it turns out, in the original tale, Cinderella did not wear glass slippers, but instead wore squirrel slippers. The explanation to this conundrum lies in the similarity of two French words, "verre," which means glass and "vair," which means squirrel. The white belly hair of the Russian squirrels (*Sciurus vulgaris*) commonly was used to decorate the clothing of royalty. (White furs of other animals were also used.) Presumably, when the tale of Cinderella finally was written down, the fur ("vair") slippers were transmuted into glass ("verre") slippers, surely much to the discomfort of Cinderella.

What roles do squirrels play in popular culture?

Squirrels have figured prominently in popular culture, including television, movies, and advertising campaigns. Tufty, a Eurasian red squirrel, figured prominently in a 1960s advertising campaign by the British Society for the Prevention of Accidents. Cartoons featuring Tufty instructed children on how to safely cross the street. Other famous cartoon squirrels include the Disney chipmunks, Chip and Dale; the Hanna Barbera sleuth, Secret Squirrel; and Rocky the flying squirrel, the other half of the famous Rocky and Bullwinkle duo. More recently, squirrels have shown up in television commercials selling products from nut-laden cereal to car insurance. Squirrels have also figured as mascots for sports teams and logos for companies.

How are squirrels incorporated into poetry?

Like a small grey
coffee-pot,
sits the squirrel.

Humbert Wolfe, "The Grey Squirrel"

We were surprised to discover how frequently poets have mentioned squirrels. In some cases, the purpose is didactic. The small size of the squirrel is utilized to convey a message like that in the Ramayana—each according to his own abilities—in Ralph Waldo Emerson's poem "The Squirrel and the Mountain"—and in another case, the independence of squirrels is extolled in a delightful poem by William Butler Yeats. Both poems are short, so we include them in their entirety.

The Mountain and the squirrel
Had a quarrel;
And the former called the latter "Little Prig."
Bun replied,
"You are doubtless very big;
But all sorts of things and weather
Must be taken in together,
To make up a year
And a sphere.
And I think it no disgrace
To occupy my place.
If I'm not as large as you,
You are not so small as I,

And not half so spry.
I'll not deny you make
A very pretty squirrel track;
Talents differ; all is well and wisely put;
If I cannot carry forests on my back,
Neither can you crack a nut.

RALPH WALDO EMERSON, "Fable"

Being out of heart with government
I took a broken root to fling
Where the proud, wayward squirrel went,
Taking delight that he could spring;
And he, with that low whinnying sound
That is like laughter, sprang again
And so to the other tree at a bound.
Nor the tame will, nor timid brain,
Bred that fierce tooth and cleanly limb
And threw him up to laugh on the bough;
No government appointed him.

WILLIAM BUTLER YEATS, "An Appointment"

We were also surprised at the absence of poems extolling the industrious behavior of squirrels collecting and storing nuts for the winter. Perhaps there are such poems but they are not memorable. Instead, poets seemed to use squirrels to elicit moods and to set the time of year. We give several examples here. These are only fragments of each poem.

The squirrel gloats on his accomplish'd hoard,
The ants have brimm'd their garners with ripe grain,
And honey bees have stor'd
The sweets of Summer in their luscious cells;
The swallows all have wing'd across the main;
But here the Autumn melancholy dwells,
And sighs her tearful spells . . .

THOMAS HOOD, "Ode—Autumn"

And now, when comes the calm mild day, as still such days will come,
To call the squirrel and the bee from out their winter home;
When the sound of dropping nuts is heard, though all the trees are still,
And twinkle in the smoky light the waters of the rill,

The south wind searches for the flowers whose fragrance late he bore,
And sighs to find them in the wood and by the stream no more.

WILLIAM CULLEN BRYANT, "The Death of the Flowers"

O what can ail thee, knight-at-arms,
Alone and palely loitering?
The sedge has wither'd from the lake,
And no birds sing.
'O what can ail thee, knight-at-arms!
So haggard and so woe-begone?
The squirrel's granary is full,
And the harvest's done.

JOHN KEATS, "La Belle Dame Sans Merci"

Some poets have taken a naturalistic viewpoint and have described straightforward observations of squirrels, such as the shyness of squirrels in hunted populations, described by Yeats, the vulnerability of a larder hoard to the depredations of the grizzly, described by Bret Harte; or the scolding of the North American red squirrel, described by Ernest Crosby; or that of the Indian striped squirrel, noted by Rudyard Kipling.

Come play with me;
Why should you run
Through the shaking tree
As though I'd a gun
To strike you dead?
When all I would do
Is to scratch your head
And let you go

WILLIAM BUTLER YEATS, "To a Squirrel at Kyle-na-gno"

Coward,—of heroic size,
In whose lazy muscles lies
Strength we fear and yet despise;
Savage,—whose relentless tusks
Are content with acorn husks;
Robber,—whose exploits ne'er soared
O'er the bee's or squirrel's hoard;
Whiskered chin, and feeble nose,
Claws of steel on baby toes,—
Here, in solitude and shade,

Shambling, shuffling plantigrade,
Be thy courses undismayed!

Francis Bret Harte, "Grizzly"

As I sit on a log here in the woods among the clean-faced beeches,
The trunks of the trees seem to me like the pipes of a mighty organ,
Thrilling my soul with wave on wave of the harmonies of the universal
    anthem—
The grand, divine, eonic "I am" chorus.

The red squirrel scolding in yonder hickory tree,
The flock of blackbirds chattering in council overhead,
The monotonous crickets in the unseen meadow,
Even the silent ants traveling their narrow highway with enormous
    burdens at my feet—
All, like choristers, sing in the green-arched cathedral
The heaven-prompted mystery, "I am, I am."

Ernest Crosby, "Choir Practice"

My garden blazes brightly with the rose-bush and the peach,
And the köil sings above it, in the siris by the well,
From the creeper-covered trellis comes the squirrel's chattering speech,
And the blue jay screams and flutters where the cheery sat-bhai dwell.

Rudyard Kipling, "In Springtime"

In other poems, squirrels are used in simile or metaphor to describe human characteristics. George Meredith's simile may seem strange to urban or suburban residents who see unhunted squirrels, but it will seem very apt for those who see squirrels in hunted woods. Yeats use of the phrase, "he had a squirrel's eye", strikes us as peculiar, but may help more knowledgeable readers to recognize Aengus in his disguise.

Shy as the squirrel and wayward as the swallow,
Swift as the swallow along the river's light
Circleting the surface to meet his mirror'd winglets,
Fleeter she seems in her stay than in her flight.
Shy as the squirrel that leaps among the pine-tops,
Wayward as the swallow overhead at set of sun,
She whom I love is hard to catch and conquer,
Hard, but O the glory of the winning were she won!

George Meredith, "Love in the Valley"

Squirrels in Stories and Literature

But, more than all, his heart is stung
To think of one, almost a child;
A sweet and playful Highland girl,
As light and beauteous as a squirrel,
As beauteous and as wild!

WILLIAM WORDSWORTH, "Peter Bell—A Tale"

He had ragged long grass-coloured hair;
He had knees that stuck out of his hose;
He had puddle-water in his shoes;
He had half a cloak to keep him dry,
Although he had a squirrel's eye.

WILLIAM BUTLER YEATS, "Baile and Aillinn"

How are squirrels incorporated into literature?

Squirrels enter into all sorts of literature—children's, adults', fiction, and nonfiction. In children's literature, an excellent example is *The Tale of Squirrel Nutkin*, by Beatrix Potter, published in 1903. It featured the Eurasian red squirrel (*Sciurus vulgaris*), and it contains an interesting description of squirrels sailing, instead of swimming. There is a fine illustration of this, painted by Beatrix Potter, in which she shows a flotilla of squirrels, each floating on a little raft of twigs with its tail upright to serve as a sail, as they all cross to Owl Island to collect nuts. The tale of squirrels sailing has a long history, appearing in English in 1658 in Edward Topsell's *The History of Four-Footed Beasts and Serpents and Insects*, which in turn was based on Conrad Gesner's *Historiae Animalium* of 1602. John Burroughs was another author of children's literature. His writings tended to be based on direct observation and are more naturalistic. In his book *Squirrels and Other Fur-Bearers*, he provides descriptions of the behavior and ecology of the eastern gray squirrel, North American red squirrel, southern flying squirrel, eastern chipmunk, and woodchuck.

In 1923, Felix Salten published a book, *Bambi, eine Lebensgeschichte aus dem Walde*, translated into English in 1929, by Whittaker Chambers, and later popularized by Disney's film *Bambi*, in 1942. Salten included squirrels—the Eurasian red squirrel (*Sciurus vulgaris*)—in the story, but they played minor roles. He subsequently wrote a book, *Die Jogend des Eichornchens Perri*, about the Eurasian red squirrel, which was greatly modified and popularized by Disney in his 1957 film, *Perri*. These are only three of the many authors of children's literature who incorporated squirrels into their writings. We have selected them as being among the most famous, but others would probably make different selections.

In adult literature, we see that squirrels are incorporated in many of the same ways as in poetry. Squirrels are used didactically, in some cases by the same authors, such as Emerson. We particularly like the allusion to the way a light coat of snow reveals the movements of animals from their tracks. A child discovering this by himself will long bear wonderful memories of it. Unlike the absence of references in poetry to the hoarding of nuts, we do find Emerson remarking on it, but not to promote the lesson of thoughtfully providing for the future, which we had expected.

Even a blind squirrel finds an acorn sometimes.

PROVERB

Commit a crime and the world is made of glass. Commit a crime, and it seems as if a coat of snow fell on the ground, such as reveals in the woods the track of every partridge and fox and squirrel and mole.

RALPH WALDO EMERSON, "Compensation"

The squirrel hoards nuts and the bee gathers honey, without knowing what they do, and they are thus provided for without selfishness or disgrace.

RALPH WALDO EMERSON, "The Transcendentalist"

In many cases, squirrels are described in such a way as to epitomize the whole of nature or at least some aspect of it. As such, they demonstrate some attribute of squirrels that was greatly admired by the author. Our lead contributor is again Emerson.

A squirrel leaping from bough to bough, and making the wood but one wide tree for his pleasure, fills the eye not less than a lion,—is beautiful, self-sufficing, and stands then and there for nature.

RALPH WALDO EMERSON, "Art," Essays, First Series (1841, repr. 1847)

The wildness of squirrels is an awesome wildness.

DOUGLAS FAIRBAIRN

That author who draws a character, even though to common view incongruous in its parts, as the flying-squirrel, and, at different periods, as much at variance with itself as the caterpillar is with the butterfly into which it changes, may yet, in so doing, be not false but faithful to facts.

HERMAN MELVILLE, *The Confidence-Man* (1857)

If we had a keen vision and feeling of all ordinary human life, it would be like hearing the grass grow and the squirrel's heart beat, and we should die of that roar which lies on the other side of silence. As it is, the best of us walk about well wadded with stupidity.

GEORGE ELIOT [Mary Ann (or Marian) Evans], *Middlemarch*, bk. 2, ch. 20

We conclude with naturalistic presentations, by authors who obviously had watched squirrels carefully and knew their behavior and their ecology. Thoreau is our prime example, and the first quote from *Walden*, is our favorite.

I am on the alert for the first signs of spring, to hear the chance note of some arriving bird, or the striped squirrel's chirp, for his stores must be now nearly exhausted, or see the woodchuck venture out of his winter quarters.

HENRY DAVID THOREAU, *Walden* (1854)

We also saw and heard several times the red squirrel. . . . This, according to the Indian, is the only squirrel found in those woods, except a very few striped ones. It must have a solitary time in that dark evergreen forest, where there is so little life, seventy-five miles from a road as we had come. I wondered how he could call any particular tree there his home; and yet he would run up the stem of one out of the myriads, as if it were an old road to him. How can a hawk ever find him there? I fancied that he must be glad to see us, though he did seem to chide us."

HENRY DAVID THOREAU, "The Allegash and East Branch,"
in *The Maine Woods*

In almost every wood you will see where the red or gray squirrels have pawed down through the snow in a hundred places, sometimes two feet deep, and almost always directly to a nut or a pine cone, as directly as if they had started from it and bored upward . . . You wonder if they remember the locality of their deposit or discover it by the scent.

The red squirrel commonly has its winter abode in the earth under a thicket of evergreens, frequently under a small clump of evergreens in the midst of a deciduous wood.

HENRY DAVID THOREAU, "The Dispersion of Seeds,"
in *Faith in a Seed*

Occasionally a chickaree or chipmunk scurried out from among the trunks of the great pines to pick up the cones which he had previously bitten off from the upper branches; a noisy Clarke's crow clung for some time in the top of a hemlock; and occasionally flocks of cross-bill went by, with swift undulating flight and low calls.

THEODORE ROOSEVELT, *Ranch Life and the Hunting-Trail* (1896)
Chapter 12: The Game of the High Peaks: The White Goat

All these I was able to show John Burroughs when he came to visit us; although, by the way, he did not appreciate as much as we did one set of inmates of the cottage—the flying squirrels. We loved having the flying squirrels, father and mother and half-grown young, in their nest among the rafters; and at night we slept so soundly that we did not in the least mind the wild gambols of the little fellows through the rooms, even when, as sometimes happened, they would swoop down to the bed and scuttle across it.

THEODORE ROOSEVELT, *An Autobiography* (1913)
Chapter 9: Outdoors and Indoors

Chapter 12

"Squirrelology"

Which species are best known?

A great deal of information is available about the common North American squirrels. The eastern gray squirrel and the eastern fox squirrel are game animals, hunted in many areas, so a large literature exists on both species, dealing with game management. There is a whole book by Durward Allen, *Michigan Fox Squirrel Management*, for example, and the *Journal of Wildlife Management* includes many articles about the management of eastern gray squirrel populations. In addition, many biologists have studied them, leading to books like *The World of the Gray Squirrel*, by Fred Barkalow and Monica Shorten, and *North American Tree Squirrels*, by Michael Steele and John Koprowski. Squirrels are very poor choices as laboratory animals, so most of the literature deals with studies in the wild.

The North American red squirrel (*Tamiasciurus hudsonicus*) has also been studied extensively—Michael Steele's review of its biology in 1998 listed three full pages of references to studies of this species. The North American red squirrel has long been known to store food in larders and to defend these aggressively, but until a paper by Fritz Gerhardt was published in 2005, we were unaware how important pilfering is in this species. This should be a lesson to us all that there are probably large gaps in our knowledge, even of the most well-known animals.

The Eurasian red squirrel (*Sciurus vulgaris*) has been the subject of many studies, too, most recently because introductions of the eastern gray squirrel have extirpated the Eurasian red squirrel over large areas (as described in our question about invasive species). John Gurnell's book, *The Natural*

A yellow-bellied marmot (*Marmota flaviventris*). Photo © National Park Service

History of Squirrels, is a good introduction to the basic biology and ecology of the species.

The North American ground squirrels also are well known. Because many species of these ground squirrels are social, their wild, social colonies are popular study sites. For example, there is a book called *The Black-Tailed Prairie Dog* by John Hoogland, another entitled *Marmots: Social Behavior and Ecology* by David Barash, an extensive series of papers on the yellow-bellied marmot by Ken Armitage, and many studies on Richardson's ground squirrels, Columbian ground squirrels, Belding's ground squirrels, and eastern chipmunks by diverse authors. Because North American ground squirrels hibernate, several species have been maintained in laboratories and studied intensively. Accordingly, we know much more about ground squirrel physiology than that of tree squirrels.

Some African and Southeast Asian squirrels have been well studied. The giant tree squirrels of southern Asia have been the subjects of extended studies by biologists Junaidi (formerly John) Payne in Malaysia (*Ratufa bicolor* and *R. affinis*) and Renee Borges in India (*Ratufa indica*). Smithsonian researcher Louise Emmons conducted an extensive study of nine species of squirrels from five genera in Gabon, Africa (*Protoxerus, Epixerus, Heliosciurus, Paraxerus,* and *Funisciurus*), and University of Pretoria professor San Viljoen studied species of *Paraxerus* and *Funisciurus* in southern Africa. Currently, Jane Waterman, of the University of Central Florida, is conducting long-term studies of the South African ground squirrel (*Xerus inauris*) and its remarkable social system.

The anatomy of squirrels is fairly well documented, initially by an excellent nineteenth-century German monograph on the anatomy of the Eurasian red squirrel. The North American squirrels—their bones and mus-

cles—were studied in detail by Monroe Bryant (1945), and a survey of the cranial osteology of all genera of squirrels was conducted by Joseph Moore (1959). A book on the anatomy of the woodchuck by Abraham Bezuidenhout and Howard Evans (2005) appeared recently. A series of shorter papers has supplemented this information in more recent years, on myology, brain size, osteology, and other features, but there is still much to be studied.

Which species are least known?

Flying squirrels are surely the least known squirrels, because they are active at night, they can move considerable distances, quickly and silently, and most species occur in Southeast Asia. The two North American species (*Glaucomys volans* and *Glaucomys sabrinus*), the Eurasian species (*Pteromys volans*), and the Japanese giant flying squirrel (*Petaurista leucogenys*) are the best known, but we have only tantalizing tidbits of knowledge about most others. Many of the Southeast Asian flying squirrels are known only from museum studies of skins and skeletal material. A wide variety of tantalizing questions remain. How does the very large woolly flying squirrel (*Eupetaurus cinereus*), which lives in caves in northern Pakistan and Afghanistan, survive on a diet of pine needles, as reported by Peter Zahler? Why does the complex-toothed flying squirrel of southern China, which also lives in caves, have such bizarre teeth? Do the diminutive Selangor pygmy flying squirrels (*Petaurillus kinlochii*) signal to one another with their feather-like, white-tipped tails?

The South American tree squirrels are also poorly known. Two Panamanian species, the red-tailed squirrel (*Sciurus granatensis*) and the variegated squirrel (*Sciurus variegatoides*), have been studied in the wild, and more limited studies of other species have been conducted, but most of the

Squirrels: The Animal Answer Guide

A juvenile eastern fox squirrel (*Sciurus niger*) inspects the photographer. Photo © Gregg Elovich, www .scarysquirrel.org

South American species of *Sciurus* and *Microsciurus* are effectively unstudied. Are the large species specialists on hard palm nuts, like the red-tailed squirrel in Panama? Are the small species all bark gleaners?

In Africa, Emmons's and Viljoen's studies of five species of rope squirrels (*Funisciurus*) demonstrate that considerable differences in behavior and habitat exist. The other species are unstudied and deserve attention. Is the fire-footed rope squirrel (*Funisciurus pyrropus*) the only species that nests in terrestrial burrows? Do different species in different parts of Africa exhibit similar behaviors and ecologies? The sun squirrels (*Heliosciurus*) also are relatively unstudied. Do all the forms of the Gambian sun squirrel (*Heliosciurus gambianus*) throughout the dry forests of Africa exhibit the same behavior and ecology? Is this true of the red-legged sun squirrel of the wet high forest (*Heliosciurus rufobrachium*) and the closely related mutable sun squirrel (*Heliosciurus mutabilis*)? Detailed studies of the behavior and ecology of the diverse species of the bush squirrels, *Paraxerus*, also are bound to be interesting.

In southern Asia, there are clearly fascinating things to be learned about the biology of squirrels, as exemplified by the recent surprising report of nectar robbing by Swinhoe's striped squirrel (*Tamiops swinhoei*). In India, the two species of palm squirrel, the Indian palm squirrel (*Funambulus palmarum*) and the northern palm squirrel (*Funambulus pennanti*), beckon for a comparative study. Do the diverse species of the *Callosciurus* tree squirrels differ as much in their behavior and ecology as they do in their coat colors and patterns? How do the ecologies and behaviors of the southern Asian ground squirrels (*Lariscus*, *Menetes*, and *Rhinosciurus*) differ from those of the tree shrews (*Tupaia*), which look similar and live in similar habitats?

How do scientists recognize individual squirrels?

Biologists who wish to study squirrels in the wild first usually capture the animals in "live traps," mark each individual in a unique way, and release them again. This enables the researcher to recognize individuals and, in long-term studies, to determine the ages and social relationships of the individuals. For short-term marking, they may write a number on the sides of the animals with hair dye, which will last until the next molt. Ball-chain collars with colored beads on them are also used and may last longer than the hair dye. Numbered ear tags are more permanent, but they require that the animals be recaptured for identification, unless the ear tags have been painted in a distinct way that is observable from a distance. There are many other ways to mark squirrels for identification, including the use of radio collars that enable you to locate the animals at any time of night or day. Another technique involves the use of pit tags, which are small transponders that are injected under the skin and will send a unique signal upon receiving particular radio waves. These are useful for close-range monitoring of animal activity, such as animals moving into or out of a nest hole, but they are not useful at longer distances. In summary, no one way seems to be perfect or infallible, and some ways of marking are better than others for particular kinds of study. In our studies, we have combined several marking techniques, such as ball-chain collars and ear tags, with good success.

Squirrels: The Animal Answer Guide

Appendix

Squirrels of the World

Order Rodentia
Family Sciuridae

Scientific Name	Common Name	General Location
Subfamily Ratufinae		
Ratufa affinis	pale giant squirrel	Southeast Asia
Ratufa bicolor	black giant squirrel	Southeast Asia
Ratufa indica	Indian giant squirrel	South Asia
Ratufa macroura	Sri Lankan giant squirrel	South Asia
Subfamily Sciurillinae		
Sciurillus pusillus	Neotropical pygmy squirrel	South America
Subfamily Sciurinae		
Tribe Sciurini		
Microsciurus alfari	Central American dwarf squirrel	Central and South America
Microsciurus flaviventer	Amazon dwarf squirrel	South America
Microsciurus mimulus	western dwarf squirrel	Central and South America
Microsciurus santanderensis	Santander dwarf squirrel	South America
Rheithrosciurus macrotis	tufted ground squirrel	Southeast Asia
Sciurus aberti	Abert's squirrel	North America and Mexico
Sciurus aestuans	Guianan squirrel	South America
Sciurus alleni	Allen's squirrel	Mexico
Sciurus anomalus	Caucasian squirrel	North Asia
Sciurus arizonensis	Arizona gray squirrel	North America and Mexico
Sciurus aureogaster	red-bellied squirrel	Mexico
Sciurus carolinensis	eastern gray squirrel	North America (introduced Europe)
Sciurus colliaei	Collie's squirrel	Mexico
Sciurus deppei	Deppe's squirrel	Mexico and Central America

Sciurus flammifer	fiery squirrel	South America
Sciurus gilvigularis	yellow-throated squirrel	South America
Sciurus granatensis	red-tailed squirrel	Central and South America
Sciurus griseus	western gray squirrel	Mexico and North America
Sciurus ignitus	Bolivian squirrel	South America
Sciurus igniventris	Northern Amazon red squirrel	South America
Sciurus lis	Japanese squirrel	North Asia
Sciurus nayaritensis	Mexican fox squirrel	Mexico and North America
Sciurus niger	eastern fox squirrel	North America
Sciurus oculatus	Peters's squirrel	Mexico
Sciurus pucheranii	Andean squirrel	South America
Sciurus pyrrhinus	Junín red squirrel	South America
Sciurus richmondi	Richmond's squirrel	Central America
Sciurus sanborni	Sanborn's squirrel	South America
Sciurus spadiceus	Southern Amazon red squirrel	South America
Sciurus stramineus	Guayaquil squirrel	South America
Sciurus variegatoides	variegated squirrel	Mexico and Central America
Sciurus vulgaris	Eurasian red squirrel	Europe and North Asia
Sciurus yucatanensis	Yucatan squirrel	Mexico and Central America
Syntheosciurus brochus	Bangs's mountain squirrel	Central America
Tamiasciurus douglasii	Douglas' squirrel	North America
Tamiasciurus hudsonicus	red squirrel	North America
Tamiasciurus mearnsi	Mearns's squirrel	Mexico
Tribe Pteryomini		
Aeretes melanopterus	Northern Chinese flying squirrel	North Asia
Aeromys tephromelas	black flying squirrel	Southeast Asia
Aeromys thomasi	Thomas's flying squirrel	Southeast Asia
Belomys pearsonii	hairy-footed flying squirrel	South Asia
Biswamoyopterus biswasi	Namdapha flying squirrel	South Asia
Eoglaucomys fimbriatus	Kashmir flying squirrel	South Asia
Eupetaurus cinereus	woolly flying squirrel	South Asia
Glaucomys sabrinus	northern flying squirrel	North America
Glaucomys volans	southern flying squirrel	North America
Hylopetes alboniger	particolored flying squirrel	Southeast Asia
Hylopetes bartelsi	Bartel's flying squirrel	Southeast Asia
Hylopetes lepidus	gray-cheeked flying squirrel	Southeast Asia

Hylopetes nigripes	Palawan flying squirrel	Southeast Asia
Hylopetes phayrei	Indochinese flying squirrel	Southeast Asia
Hylopetes platyurus	Jentink's flying squirrel	Southeast Asia
Hylopetes sipora	Sipora flying squirrel	Southeast Asia
Hylopetes spadiceus	red-cheeked flying squirrel	Southeast Asia
Hylopetes winstoni	Sumatran flying squirrel	Southeast Asia
Iomys horsfieldii	Javanese flying squirrel	Southeast Asia
Iomys sipora	Mentawai flying squirrel	Southeast Asia
Petaurillus emiliae	lesser pygmy flying squirrel	Southeast Asia
Petaurillus hosei	Hose's pygmy flying squirrel	Southeast Asia
Petaurillus kinlochii	Selangor pygmy flying squirrel	Southeast Asia
Petaurista alborufus	red and white giant flying squirrel	North and South Asia
Petaurista elegans	spotted giant flying squirrel	Southeast Asia
Petaurista leucogenys	Japanese giant flying squirrel	North Asia
Petaurista magnificus	Hodgson's giant flying squirrel	South Asia
Petaurista nobilis	Bhutan giant flying squirrel	South Asia
Petaurista petaurista	red giant flying squirrel	Southeast Asia
Petaurista philippensis	Indian giant flying squirrel	South Asia
Petaurista xanthotis	Chinese giant flying squirrel	North Asia
Petinomys crinitus	Basilan flying squirrel	Southeast Asia
Petinomys fuscocapillus	Travancore flying squirrel	South Asia
Petinomys genibarbis	whiskered flying squirrel	Southeast Asia
Petinomys hageni	Hagen's flying squirrel	Southeast Asia
Petinomys lugens	Sipora flying squirrel	Southeast Asia
Petinomys mindanensis	Mindanao flying squirrel	Southeast Asia
Petinomys sagitta	arrow flying squirrel	Southeast Asia
Petinomys setosus	Temminck's flying squirrel	Southeast Asia
Petinomys vordermanni	Vordermann's flying squirrel	Southeast Asia
Pteromys momonga	Japanese flying squirrel	North Asia
Pteromys volans	Siberian flying squirrel	Europe and North Asia
Pteromyscus pulverulentus	smoky flying squirrel	Southeast Asia
Trogopterus xanthipes	complex-toothed flying squirrel	North Asia

Subfamily Callosciurinae (South and Southeast Asia)

Callosciurus adamsi	ear-spot squirrel	Southeast Asia
Callosciurus albescens	Kloss's squirrel	Southeast Asia
Callosciurus baluensis	Kinabalu squirrel	Southeast Asia
Callosciurus caniceps	gray-bellied squirrel	Southeast Asia
Callosciurus erythraeus	Pallas's squirrel	Southeast Asia
Callosciurus finlaysonii	Finlayson's squirrel	Southeast Asia

Squirrels of the World

Callosciurus inornatus	inornate squirrel	Southeast Asia
Callosciurus melanogaster	Mentawai squirrel	Southeast Asia
Callosciurus nigrovittatus	black-striped squirrel	Southeast Asia
Callosciurus notatus	plantain squirrel	Southeast Asia
Callosciurus orestes	Borneo black-banded squirrel	Southeast Asia
Callosciurus phayrei	Phayre's squirrel	Southeast Asia
Callosciurus prevostii	Prevost's squirrel	Southeast Asia
Callosciurus pygerythrus	Irrawaddy squirrel	South Asia
Callosciurus quinquestriatus	Anderson's squirrel	South Asia
Dremomys everetti	Bornean Mountain ground squirrel	Southeast Asia
Dremomys gularis	red-throated squirrel	South Asia
Dremomys lokriah	orange-bellied Himalayan squirrel	South Asia
Dremomys pernyi	Perny's long-nosed squirrel	South Asia
Dremomys pyrrhomerus	red-hipped squirrel	South Asia
Dremomys rufigenis	Asian red-cheeked squirrel	Southeast Asia
Exilisciurus concinnus	Philippine pygmy squirrel	Southeast Asia
Exilisciurus exilis	least pygmy squirrel	Southeast Asia
Exilisciurus whiteheadi	tufted pygmy squirrel	Southeast Asia
Funambulus layardi	Layard's palm squirrel	South Asia
Funambulus palmarum	Indian palm squirrel	South Asia
Funambulus pennantii	northern palm squirrel	South Asia
Funambulus sublineatus	dusky palm squirrel	South Asia
Funambulus tristriatus	jungle palm squirrel	South Asia
Glyphotes simus	sculptor squirrel	Southeast Asia
Hyosciurus heinrichi	montane long-nosed squirrel	Southeast Asia
Hyosciurus ileile	lowland long-nosed squirrel	Southeast Asia
Lariscus hosei	four-striped ground squirrel	Southeast Asia
Lariscus insignis	three-striped ground squirrel	Southeast Asia
Lariscus niobe	Niobe ground squirrel	Southeast Asia
Lariscus obscurus	Mentawai three-striped squirrel	Southeast Asia
Menetes berdmorei	Indochinese ground squirrel	Southeast Asia
Nannosciurus melanotis	black-eared squirrel	Southeast Asia
Prosciurillus abstrusus	secretive dwarf squirrel	Southeast Asia
Prosciurillus leucomus	whitish dwarf squirrel	Southeast Asia
Prosciurillus murinus	Celebes dwarf squirrel	Southeast Asia
Prosciurillus rosenbergii	Sanghir squirrel	Southeast Asia
Prosciurillus weberi	Weber's dwarf squirrel	Southeast Asia
Rhinosciurus laticaudatus	shrew-faced squirrel	Southeast Asia
Rubrisciurus rubriventer	Sulawesi giant squirrel	Southeast Asia
Sundasciurus brookei	Brooke's squirrel	Southeast Asia

Sundasciurus davensis	Davao squirrel	Southeast Asia
Sundasciurus fraterculus	fraternal squirrel	Southeast Asia
Sundasciurus hippurus	horse-tailed squirrel	Southeast Asia
Sundasciurus hoogstraali	Busuanga squirrel	Southeast Asia
Sundasciurus jentinki	Jentinck's squirrel	Southeast Asia
Sundasciurus juvencus	Northern Palawan tree squirrel	Southeast Asia
Sundasciurus lowii	Low's squirrel	Southeast Asia
Sundasciurus mindanensis	Mindanao squirrel	Southeast Asia
Sundasciurus moellendorffi	Culion tree squirrel	Southeast Asia
Sundasciurus philippinensis	Philippine tree squirrel	Southeast Asia
Sundasciurus rabori	Palawan Montane squirrel	Southeast Asia
Sundasciurus samarensis	Samar squirrel	Southeast Asia
Sundasciurus steerii	Southern Palawan tree squirrel	Southeast Asia
Sundasciurus tenuis	slender squirrel	Southeast Asia
Tamiops mcclellandii	Himalayan striped squirrel	Southeast Asia
Tamiops maritimus	Maritime striped squirrel	North and Southeast Asia
Tamiops rodolphii	Cambodian striped squirrel	Southeast Asia
Tamiops swinhoei	Swinhoe's striped squirrel	North and Southeast Asia

Subfamily Xerinae
Tribe Xerini

Atlantoxerus getulus	Barbary ground squirrel	Africa
Spermophilopsis leptodactylus	long-clawed ground squirrel	Africa
Xerus erythropus	striped ground squirrel	Africa
Xerus inauris	South African ground squirrel	Africa
Xerus princeps	Damara ground squirrel	Africa
Xerus rutilus	unstriped ground squirrel	Africa

Tribe Protoxerini

Epixerus ebii	western palm squirrel	Africa
Funisciurus anerythrus	Thomas's rope squirrel	Africa
Funisciurus bayonii	Lunda rope squirrel	Africa
Funisciurus carruthersi	Carruther's mountain squirrel	Africa
Funisciurus congicus	Congo rope squirrel	Africa
Funisciurus isabella	Lady Burton's rope squirrel	Africa
Funisciurus lemniscatus	ribboned rope squirrel	Africa
Funisciurus leucogenys	red-cheeked rope squirrel	Africa
Funisciurus pyrropus	fire-footed rope squirrel	Africa
Funisciurus substriatus	Kintampo rope squirrel	Africa
Heliosciurus gambianus	Gambian sun squirrel	Africa

Heliosciurus mutabilis	mutable sun squirrel	Africa
Heliosciurus punctatus	small sun squirrel	Africa
Heliosciurus rufobrachium	red-legged sun squirrel	Africa
Heliosciurus ruwenzorii	Ruwenzori sun squirrel	Africa
Heliosciurus undulatus	Zanj sun squirrel	Africa
Myosciurus pumilio	African pygmy squirrel	Africa
Paraxerus alexandri	Alexander's bush squirrel	Africa
Paraxerus boehmi	Boehm's bush squirrel	Africa
Paraxerus cepapi	Smith's bush squirrel	Africa
Paraxerus cooperi	Cooper's mountain squirrel	Africa
Paraxerus flavovittis	striped bush squirrel	Africa
Paraxerus lucifer	black and red bush squirrel	Africa
Paraxerus ochraceus	ochre bush squirrel	Africa
Paraxerus palliatus	red bush squirrel	Africa
Paraxerus poensis	green bush squirrel	Africa
Paraxerus vexillarius	Swynnerton's bush squirrel	Africa
Paraxerus vincenti	Vincent's bush squirrel	Africa
Protoxerus aubinnii	slender-tailed squirrel	Africa
Protoxerus stangeri	forest giant squirrel	Africa
Tribe Marmotini		
Ammospermophilus harrisii	Harris's antelope squirrel	North America and Mexico
Ammospermophilus insularis	Espiritu Santo Island antelope squirrel	Mexico
Ammospermophilus interpres	Texas antelope squirrel	North America and Mexico
Ammospermophilus leucurus	white-tailed antelope squirrel	North America and Mexico
Ammospermophilus nelsoni	Nelson's antelope squirrel	North America
Cynomys gunnisoni	Gunnison's prairie dog	North America
Cynomys leucurus	white-tailed prairie dog	North America
Cynomys ludovicianus	black-tailed prairie dog	North America and Mexico
Cynomys mexicanus	Mexican prairie dog	Mexico
Cynomys parvidens	Utah prairie dog	North America
Marmota baibacina	gray marmot	North Asia
Marmota bobak	bobak marmot	Europe and North Asia
Marmota broweri	Alaska marmot	North America
Marmota caligata	hoary marmot	North America
Marmota camtschatica	black-capped marmot	North Asia
Marmota caudata	long-tailed marmot	North Asia
Marmota flaviventris	yellow-bellied marmot	North America
Marmota himalayana	Himalayan marmot	North Asia

Marmota marmota	alpine marmot	Europe
Marmota menzbieri	Menzbier's marmot	North Asia
Marmota monax	woodchuck	North America
Marmota olympus	Olympic marmot	North America
Marmota sibirica	Tarbagan marmot	North Asia
Marmota vancouverensis	Vancouver Island marmot	North America
Sciurotamias davidianus	Père David's rock squirrel	North Asia
Sciurotamias forresti	Forrest's rock squirrel	South Asia
Spermophilus adocetus	tropical ground squirrel	Mexico
Spermophilus alashanicus	Alashan ground squirrel	North Asia
Spermophilus annulatus	ring-tailed ground squirrel	Mexico
Spermophilus armatus	Uinta ground squirrel	North America
Spermophilus atricapillus	Baja California rock squirrel	Mexico
Spermophilus beecheyi	California ground squirrel	North America
Spermophilus beldingi	Belding's ground squirrel	North America
Spermophilus brevicauda	Brandt's ground squirrel	North Asia
Spermophilus brunneus	Idaho ground squirrel	North America
Spermophilus canus	Merriam's ground squirrel	North America
Spermophilus citellus	European ground squirrel	Europe
Spermophilus columbianus	Columbian ground squirrel	North America
Spermophilus dauricus	Daurian ground squirrel	North Asia
Spermophilus elegans	Wyoming ground squirrel	North America
Spermophilus erythrogenys	red-cheeked ground squirrel	North Asia
Spermophilus franklinii	Franklin's ground squirrel	North America
Spermophilus fulvus	yellow ground squirrel	North Asia
Spermophilus lateralis	golden-mantled ground squirrel	North America
Spermophilus madrensis	Sierra Madre ground squirrel	Mexico
Spermophilus major	russet ground squirrel	North Asia
Spermophilus mexicanus	Mexican ground squirrel	North America and Mexico
Spermophilus mohavensis	Mohave ground squirrel	North America
Spermophilus mollis	Piute ground squirrel	North America
Spermophilus musicus	Caucasian mountain ground squirrel	North Asia
Spermophilus pallidicauda	pallid ground squirrel	North Asia
Spermophilus parryii	Arctic ground squirrel	North America and North Asia
Spermophilus perotensis	Perote ground squirrel	Mexico
Spermophilus pygmaeus	little ground squirrel	North Asia
Spermophilus ralli	Tien Shan ground squirrel	North Asia
Spermophilus relictus	relict ground squirrel	North Asia
Spermophilus richardsonii	Richardson's ground squirrel	North America

Spermophilus saturatus	cascade golden-mantled ground squirrel	North America
Spermophilus spilosoma	spotted ground squirrel	North America and Mexico
Spermophilus suslicus	speckled ground squirrel	Europe
Spermophilus tereticaudus	round-tailed ground squirrel	North America and Mexico
Spermophilus townsendii	Townsend's ground squirrel	North America
Spermophilus tridecemlineatus	thirteen-lined ground squirrel	North America
Spermophilus undulatus	long-tailed ground squirrel	North Asia
Spermophilus variegatus	rock squirrel	North America and Mexico
Spermophilus washingtoni	Washington ground squirrel	North America
Spermophilus xanthoprymnus	Asia Minor ground squirrel	North Asia
Tamias alpinus	alpine chipmunk	North America
Tamias amoenus	yellow-pine chipmunk	North America
Tamias bulleri	Buller's chipmunk	North America and Mexico
Tamias canipes	gray-footed chipmunk	North America
Tamias cinereicollis	gray-collared chipmunk	North America
Tamias dorsalis	cliff chipmunk	North America and Mexico
Tamias durangae	Durango chipmunk	Mexico
Tamias merriami	Merriam's chipmunk	North America and Mexico
Tamias minimus	least chipmunk	North America
Tamias obscurus	California chipmunk	North America and Mexico
Tamias ochrogenys	yellow-cheeked chipmunk	North America
Tamias palmeri	Palmer's chipmunk	North America
Tamias panamintinus	Panamint chipmunk	North America
Tamias quadrimaculatus	long-eared chipmunk	North America
Tamias quadrivittatus	Colorado chipmunk	North America
Tamias ruficaudus	red-tailed chipmunk	North America
Tamias rufus	Hopi chipmunk	North America
Tamias senex	shadow chipmunk	North America
Tamias sibiricus	Siberian chipmunk	Europe and North Asia
Tamias siskiyou	Siskiyou chipmunk	North America
Tamias sonomae	Sonoma chipmunk	North America
Tamias speciosus	lodgepole chipmunk	North America
Tamias striatus	eastern chipmunk	North America
Tamias townsendii	Townsend's chipmunk	North America
Tamias umbrinus	Uinta chipmunk	North America

Bibliography

Adams, L. W., J. Hadidian, and V. Flyger. 2004. Movement and mortality of translocated urban-suburban grey squirrels. *Animal Welfare* 13:45–50.

Adler, B., Jr. 1988. *Outwitting Squirrels.* Chicago Review Press, Chicago.

Amori, G., and S. Gippoliti. 1995. Siberian chipmunk *Tamias sibiricus* in Italy. *Mammalia* 59 (2):288–289.

Ando, M., and S. Shiraishi. 1993. Gliding flight in the Japanese giant flying squirrel *Petaurista leucogenys. Journal of the Mammal Society of Japan* 18 (1):19–32.

Anufriev, A. I., and G. G. Arkhipov. 2004. Influence of body weight and size on the mode of wintering in hibernators of the Family Sciuridae in Northeastern Russia. *Russian Journal of Ecology* 35 (3):189–193.

Applegate, R. D., and R. C. McCord. 1974. A description of swimming in the fox squirrel. *American Midland Naturalist* 92 (1):255.

Aprile, G., and D. Chico. 1999. Nueva especies exotica de mamífero en la Argentina: la ardilla de vientre rojo (*Callosciurus erythraeus*). *Mastozoología Neotropical* 6 (1):7–14.

Armitage, K. B. 1999. Evolution of sociality in marmots. *Journal of Mammalogy* 80 (1):1–10.

Armitage, K. B., D. T. Blumstein, and B. C. Wood. 2003. Energetics of hibernating yellow-bellied marmots (*Marmota flaviventris*). *Comparative Biochemistry and Physiology Part A* 134:101–114.

Arnold, W. 1990. The evolution of marmot sociality: II. Costs and benefits of joint hibernation. *Behavioral Ecology and Sociobiology* 27:239–246.

Bailey, M. B. 1986. Every animal is the smartest: Intelligence and the ecological niche. In *Animal Intelligence: Insights into the Animal Mind*, ed. J. Hoage and L. Goldman, 105–113. Smithsonian Institution Press, Washington, DC.

Bailey, V. 1923. Mammals of the District of Columbia. *Proceedings of the Biological Society of Washington* 36:103–148.

Barash, D. P. 1989. *Marmots: Social Behavior and Ecology.* Stanford University Press, Stanford.

Barkalow, F. J., R. B. Hamilton, and R. F. Soots, Jr. 1970. The vital statistics of an unexploited gray squirrel population. *Journal of Wildlife Management* 34 (3):489–500.

Barkalow, F. J., and M. Shorten. 1973. *The World of the Gray Squirrel.* J. B. Lippincott Company, Philadelphia.

Belk, M. C., and H. D. Smith. 1991. *Ammospermophilus leucurus. Mammalian Species* 368:1–8.

Benkman, C. W. 1995. The impact of tree squirrels (*Tamiasciurus*) on limber pine seed dispersal adaptations. *Evolution* 49 (4):585–592.

Bennett, A. F., R. B. Huey, H. John-Adler, and K. A. Nagy. 1984. The parasol tail and thermoregulatory behavior of the cape ground squirrel *Xerus inauris. Physiological Zoology* 57 (1):57–62.

Bertolino, S., I. Currado, and P. J. Mazzoglio. 1999. Finlayson's (Variable) Squirrel *Callosciurus finlaysoni* in Italy. *Mammalia* 63 (4):522–525.

Bertolino, S., and P. Genovesi. 2003. Spread and attempted eradication of the grey squirrel (*Sciurus carolinensis*) in Italy, and consequences for the red squirrel (*Sciurus vulgaris*) in Eurasia. *Biological Conservation* 109: 351–358.

Best, T. L. 1995. *Spermophilus mohavensis. Mammalian Species* 509:1–7.

Bezuidenhout, A. J., and H. E. Evans. 2005. *Anatomy of the Woodchuck (Marmota monax).* American Society of Mammalogists, Lawrence, KS.

Biggins, D. E., and M. Y. Kosoy. 2001. Influences of introduced plague on North American mammals: Implications from ecology of plague in Asia. *Journal of Mammalogy.* 82 (4):906–916.

Bintz, G. L. 1984. Water balance, stress and seasonal torpor. In *The Biology of Ground-Dwelling Squirrels,* ed. J. O. Murie and G. R. Michener, 142–165. University of Nebraska Press, Lincoln.

Blumstein, D. T. 1999. Alarm calling in three species of marmots. *Behaviour* 136: 731–757.

Blumstein, D. T., and K. B. Armitage. 1998. Life history consequences of social complexity: a comparative study of ground dwelling sciurids. *Behavioral Ecology* 9 (1):8–19.

Blumstein, D. T., L. Verneyre, and J. C. Daniel. 2004. Reliability and the adaptive utility of discrimination among alarm callers. *Proceedings of the Royal Society of London B* 271:1851–1857.

Bordignon, M., T. C. C. Margarido, and R. R. Lange. 1996. Formas de abertura dos frutos de *Syagrus romanzoffiana* (Chamisso) Glassman efetuadas por *Sciurus ingrami* Thomas (Rodentia, Sciuridae). *Revista Brasileira de zoologia* 13 (4):821–828.

Borges, R. M. 1989. Resource heterogeneity and the foraging ecology of the Malabar Giant squirrel, *Ratufa indica.* Ph.D. thesis. University of Miami, Coral Gables, FL.

Bryant, A. A. 1996. Reproduction and persistence of Vancouver Island marmots (*Marmota vancouverensis*) in natural and logged habitats. *Canadian Journal of Zoology* 74 (4):678–687.

Bryant, A. A., and D. W. Janz. 1996. Distribution and abundance of Vancouver Island marmots (*Marmota vancouverensis*). *Canadian Journal of Zoology* 74 (4):667–677.

Bryant, A. A., D. W. Janz, M. C. deLaronde, and D. D. Doyle. 2002. Recent Vancouver Island marmot (*Marmota vancouverensis*) population changes. In *Holarctic Marmots as a Factor of Biodiversity,* ed. K. B. Armitage and V. U. Rumianstev, 88–100. ABF Publishing House, Moscow.

Bryant, M. D. 1945. Phylogeny of Nearctic Sciuridae. *American Midland Naturalist* 33 (2):257–390.

Buenau, K. E., and L. R. Gerber. 2004. Developing recovery and monitoring strategies for the endemic Mount Graham red squirrels (*Tamiasciurus hudsonicus grahamensis*) in Arizona. *Animal Conservation* 7:17–22.

Bumiller, E. 2003. A pest looks for, and gets, handouts. *New York Times.* 13 January.

Burke Da Silva, K., C. Mahan, and J. Da Silva. 2002. The trill of the chase: Eastern chipmunks call to warn kin. *Journal of Mammalogy* 83 (2):546–552.

Callahan, J. R. 1981. Vocalization and parental investment in female *Eutamias*. *American Midland Naturalist* 118 (6):872–875.

Chappell, M. A., and G. A. Bartholomew. 1981. Activity and thermoregulation of the antelope ground squirrel *Ammospermophilus leucurus* in winter and summer. *Physiological Zoology* 54 (2):215–223.

Childs, J. E., L. Colby, J. W. Krebs, T. Strine, M. Feller, D. Noah, C. Drenzek, J. S. Smith, and C. E. Rupprecht. 1997. Surveillance and spatiotemporal associations of rabies in rodents and lagomorphs in the United States, 1985–1994. *Journal of Wildlife Diseases* 33 (1):20–27.

Clark, M. F., and D. L. Kramer. 1994. Scatterhoarding by a larderhoarding rodent: Intraspecific variation in the hoarding behaviour of the eastern chipmunk, *Tamias striatus*. *Animal Behaviour* 48:299–308.

Coss, R. G., K. L. Gusé, N. S. Poran, and D. G. Smith. 1993. Development of antisnake defenses in California ground squirrels (*Spermophilus beecheyi*): II. Microevolutionary effects of relaxed selection from rattlesnakes. *Behaviour* 124 (1–2):137–163.

Davis, D. D. 1962. Mammals of the lowland rain-forest of north Borneo. *Bulletin of the Singapore National Museum* 31:1–129.

Davis, J. R., and T. C. Theimer. 2003. Increased lesser earless lizard (*Holbrookia maculata*) abundance on Gunnison's Prairie Dog colonies and short term responses to artificial prairie dog burrows. *American Midland Naturalist* 150:282–290.

Devenport, J. A., L. D. Luna, and L. D. Devenport. 2000. Placement, retrieval, and memory of caches by thirteen-lined ground squirrels. *Ethology* 106:171–183.

Deng, X. B., P. Y. Ren, J. Y. Gao, and Q. J. Li. The striped squirrel (*Tamiops swinhoei hainanus*) as a nectar robber of ginger (*Alpinia kwangsiensis*). *Biotropica* 36 (4):633–636.

Drew, K. L., O. Toien, P. M. Rivera, M. A. Smith, G. Perry, and M. E. Rice. 2002. Role of the antioxidant ascorbate in hibernation and warming from hibernation. *Comparative Biochemistry and Physiology Part C Toxicology and Pharmacology* 133C (4):483–492.

Elliot, L. 1978. Social behavior and foraging ecology of the eastern chipmunk (*Tamias striatus*) in the Adirondack Mountains. *Smithsonian Contributions to Zoology* 265:1–107.

Elliot, P. F. 1974. Evolutionary responses of plants to seed-eaters: pine squirrel predation on lodgepole pine. *Evolution* 28 (2):221–231.

Emamdie, D., and J. Warren. 1993. Varietal taste preference for Cacao *Theobroma cacao* L. by the Neotropical red squirrel *Sciurus granatensis* (Humboldt). *Biotropica* 25 (3):365–368.

Emmons, L. H. 1975. Ecology and behavior of African rainforest squirrels. Ph.D. thesis. Cornell University, Ithaca, NY.

Emmons, L. H. 1978. Sound Communication among African rainforest squirrels. *Zeitschrift für Tierpsychologie* 47 (1):1–49.

Emmons, L. H. 1979. Observations on litter size and development of some African rainforest squirrels. *Biotropica* 11 (3):207–213.

Emmons, L. H. 1980. Ecology and resource partitioning among nine species of African rain forest squirrels. *Ecological Monographs* 50 (1):31–54.

Emry, R. J., and R. W. Thorington. 1982. Descriptive and comparative osteology of the oldest fossil squirrel, *Protosciurus* (Rodentia: Sciuridae). *Smithsonian Contributions to Paleobiology* 47:1–34.

Ernest, K. A., and M. A. Mares. 1987. *Spermophilus tereticaudus*. *Mammalian Species* 274:1–9.

Ewer, R. F. 1964. Food burying in the African ground squirrel, *Xerus erythropus* (E. Geoff.). *Zeitschrift für Tierpsychologie* 22 (3):321–327.

Ferron, J. 1975. Solitary play of the red squirrel (*Tamiasciurus hudsonicus*). *Canadian Journal of Zoology* 53 (11):1495–1499.

Fishbein, D. B., A. J. Belotto, R. E. Pacer, J. S. Smith, W. G. Winkler, S. R. Jenkins, and K. M. Porter. 1986. Rabies in rodents and lagomorphs in the United States, 1971–1984: Increased cases in the woodchuck (*Marmota monax*) in mid-Atlantic states. *Journal of Wildlife Diseases* 22 (2):151–155.

Flyger, V. 1969. The 1968 squirrel "migration" in the eastern United States. Paper presented at the Northeast Fish and Wildlife Conference, White Sulphur Springs, West Virginia.

Fox, J. F. 1982. Adaptation of gray squirrel behavior to autumn germination by white oak acorns. *Evolution* 36 (4):800–809.

Frank, C. L. 1994. Polyunsaturate content and diet selection by ground squirrels (*Spermophilus lateralis*). *Ecology* 75 (2):458–463.

Frank, M., B. C. Reynolds, and R. K. O'Nions. 1999. Nd and Pb isotopes in Atlantic and Pacific water masses before and after closure of the Panama gateway. *Geology* 27 (12):1147–1150.

Frazer, J. G. 1922. *The Golden Bough*. Macmillan, New York.

French, A. R. 1988. The patterns of mammalian hibernation. *American Scientist* 76:569–575.

Glanz, W. E. 1984. Food and habitat use by two sympatric *Sciurus* species in central Panama. *Journal of Mammalogy* 65 (2):342–347.

Goheen, J.R., and R. K. Swihart. 2003. Food-hoarding behavior of gray squirrels and North American red squirrels in the central hardwood region: implications for forest regeneration. *Canadian Journal of Zoology* 81:1636–1639.

Gorman, R. C. 1981. *Nudes & Foods: Gorman Goes Gourmet*. Northland Press, Flagstaff, AZ.

Greene, E., and T. Meagher. 1998. Red squirrels, *Tamiasciurus hudsonicus*, produce predator-class specific alarm calls. *Animal Behaviour* 55 (3):511–518.

Gurnell, J. 1987. *The Natural History of Squirrels*. Facts on File Publications, New York.

Hadj-Chikh, L. Z., M. A. Steele, and P. D. Smallwood. 1996. Caching decisions by grey squirrels: a test of the handling time and perishability hypotheses. *Animal Behaviour* 52 (5):941–948.

Hall, J. G. 1981. A field study of the Kaibab squirrel in Grand Canyon National Park. *Wildlife Monographs* 75:1–54.

Handley, C. O. 1976. Mammals of the Smithsonian Venezuelan project. *Brigham Young University Science Bulletin, Biological Series* 20 (5):1–89.

Hanson, M. T., and R. G. Coss. 2001. Age differences in the response of California ground squirrels (*Spermophilus beecheyi*) to conspecific alarm calls. *Ethology* 107:259–275.

Hare, J. F., and B. A. Atkins. 2001. The squirrel that cried wolf: reliability detection by juvenile Richardson's ground squirrels (*Spermophilus richardsonii*). *Behavioral Ecology and Sociobiology* 51:108–112.

Hare, J. F., G. Todd, and W. A. Untereiner. 2004. Multiple mating results in multiple paternity in Richardson's ground squirrels, *Spermophilus richardsonii*. *Canadian Field Naturalist* 118 (1):90–94.

Hatt, R. T. 1929. The red squirrel: its life history and habits. *Bulletin of the New York State College of Forestry* 2 (1b):1–146.

Haynie, M. L., R. A. Van Den Bussche, J. L. Hoogland, and D. A. Gilbert. 2003. Parentage, multiple paternity, and breeding success in Gunnison's and Utah prairie dogs. *Journal of Mammalogy* 84 (4):1244–1253.

Heaney, L. R. 1995. Population vulnerability of mammals in isolated habitats. In *Storm over a Mountain Island: Conservation Biology and the Mt. Graham Affair*, ed. C. A. Istock and R. S. Hoffmann, 179–192. University of Arizona Press, Tucson.

Heffner, R. S., H. E. Heffner, C. Contos, and D. Kearns. 1994. Hearing in prairie dogs: transition between surface and subterranean rodents. *Hearing Research* 73:185–189.

Heller, H. C., X. J. Musacchia, and L. C. H. Wang, eds. 1986. *Living in the Cold: Physiological and Biochemical Adaptations*. Elsevier, New York.

Hodos, W. 1986. The evolution of the brain and the nature of animal intelligence. In *Animal Intelligence: Insights into the Animal Mind*, ed. R. J. Hoage and L. Goldman, 77–87. Smithsonian Institution Press, Washington, DC.

Hoffmann, C. K., and H. Weyenbergh, Jr. 1870. Die Osteologie und Myologie von *Sciurus Vulgaris* L., Verglichen mit der Anatomie der Lemuriden und des Chiromys und Ueber die Stellung des Letzteren im Natürlichen systeme, 136 pp. Loosjes Erben, Haarlem.

Holmes, D. J., and S. N. Austad. 1994. Fly now die later: life-history correlates of gliding and flying in mammals. *Journal of Mammalogy* 75 (1):224–226.

Hoogland, J. L. 1981. The evolution of coloniality in white-tailed and black-tailed prairie dogs (Sciuridae: *Cynomys leucurus* and *C. ludovicianus*). *Ecology* 62:252–272.

Hoogland, J. L. 1995. *The Black-Tailed Prairie Dog: Social Life of a Burrowing Mammal*. University of Chicago Press, Chicago.

Hoogland, J. L. 1996. Why do Gunnison's prairie dogs give anti-predator calls? *Animal Behaviour* 51 (4):871–880.

Hoogland, J. L. 1998. Why do female Gunnison's prairie dogs copulate with more than one male? *Animal Behaviour* 55 (2):351–359.

Hoogland, J. L. 1999. Philopatry, dispersal, and social organization of Gunnison's prairie dogs. *Journal of Mammalogy* 80 (1):243–251.

Horwich, R. H. 1972. The ontogeny of social behavior in the gray squirrel (*Sciurus carolinensis*). *Advances in Ethology* 8:1–103.

Hudson, J. W., and D. R. Deavers. 1973. Thermoregulation at high ambient temperatures of six species of ground squirrels from different habitats. *Physiological Zoology* 46 (2):95–109.

Hudson, J. W., D. R. Deavers, and S. R. Bradley. 1972. A comparative study of the temperature regulation in ground squirrels with special reference to the desert species. *Symposium of the Zoological Society of London* 31:191–213.

Inouye, D.W., B. Barr, K. B. Armitage, and B. D. Inouye. 2000. Climate change is affecting altitudinal migrants and hibernating species. *Proceedings of the National Academy of Sciences of the United States of America* 97 (4):1630–1633.

Jackson, L. L., H. E. Heffner, and R. S. Heffner. 1997. Audiogram of the fox squirrel (*Sciurus niger*). *Journal of Comparative Psychology* 111 (1):100–104.

Jacobs, L. F., and E. R. Liman. 1991. Grey squirrels remember the locations of buried nuts. *Animal Behaviour* 41 (1):103–110.

Jamieson, S. H., and K. B. Armitage. 1987. Sex differences in the play behavior of yearling yellow-bellied marmots. *Ethology* 74 (3):237–253.

Johnson, D. E., D. D. Parker, and E. D. Vest. 1968. Ecological relationships of plant communities and ectoparasites of rodents in the Great Salt Lake Desert. *Proceedings of the Utah Academy of Science, Arts and Letters* 45:130–147.

Kenward, R. E. 1983. The causes of damage by red and gray squirrels. *Mammal Review* 13 (2/3/4):159–166.

Kenward, R. E. 1990. Bark-stripping by grey squirrels in Britain and North America: Why does damage differ? In *Mammals as Pests*, ed. R. J. Putman. Chapman and Hall, New York.

Key, G. 1990. Pre-harvest crop losses to the African striped ground squirrel *Xerus erythropus* in Kenya. *Tropical Pest Management* 36 (3):223–229.

Kiltie, R. A. 1989. Wildfire and the evolution of dorsal melanism in fox squirrels. *Journal of Mammalogy* 70 (4):726–739.

Kiltie, R. A. 1992. Tests of hypotheses on predation as a factor maintaining polymorphic melanism in coastal plain fox squirrels. *Biological Journal of the Linnaean Society* 45:17–37.

King, J. A. 1955. Social behavior, social organization, and population dynamics in a black-tailed prairiedog town in the Black Hills of South Dakota. *Contributions from the Laboratory of Vertebrate Biology, The University of Michigan, no. 67*, 1–123.

Koprowski, J. L. 1993. Alternative reproductive tactics in male eastern gray squirrels: Making the best of a bad job. *Behavioral Ecology* 4 (2):165–171.

Koprowski, J. L. 1996. Natal philopatry, communal nesting and kinship in fox squirrels and gray squirrels. *Journal of Mammalogy* 77 (4):1006–1016.

Koprowski, J. L., J. L. Roseberry, and W. D. Klimstra. 1988. Longevity records for the fox squirrel. *Journal of Mammalogy* 69 (2):383–384.

Kotliar, N. B., B. W. Baker, A. D. Whicker, and G. Plumb. 1999. A critical review of assumptions about the prairie dog as a keystone species. *Environmental Management* 24 (2):177–192.

Kruckenhauser, L., and W. Pinsker. 2004. Microsatellite variation in autochthonous and introduced populations of the Alpine marmot (*Marmota marmota*) along a European west-east transect. *Journal of Zoological Systematics and Evolutionary Biology* 42:19–26.

Laundre, J. W. 1993. Effects of small mammal burrows on water infiltration in a cool desert environment. *Oecologia* 94 (1):43–48.

Layne, J. N. 1954. The biology of the red squirrel, *Tamiasciurus hudsonicus loquax* (Bangs), in Central New York. *Ecological Monographs* 24 (3):227–268.

Layne, J. N., and M. A. V. Raymond. 1994. Communal nesting of southern flying squirrels in Florida. *Journal of Mammalogy* 75:110–120.

Lehmer, E. M., B. Van Horne, B. Kulbartz, and G. L. Florant. 2001. Facultative torpor in free-ranging black-tailed prairie dogs (*Cynomys ludovicianus*). *Journal of Mammalogy* 82 (2):551–557.

Leopold, A. 1993. *Round River*, 145–146. Oxford University Press, New York.

Levenson, H. 1979. Sciurid growth rates: some corrections and additions. *Journal of Mammalogy* 60 (1):232–235.

Liat, L. B., and I. Muul. 1978. Small mammals. In *Kinabalu: Summit of Borneo*, ed. D. A. Luping, C. Wen, and E. R. Dingley, 403–457. Sabah Society, Sabah, Malaysia.

Linn, I., and G. Key. 1996. Use of space by the African striped ground squirrel *Xerus erythropus*. *Mammal Review* 26 (1):9–26.

Lishak, R. S. 1982a. Vocalization of nestling gray squirrels. *Journal of Mammalogy* 63 (3):446–452.

Lishak, R. S. 1982b. Gray squirrel mating calls: a spectrographic and ontogenic analysis. *Journal of Mammalogy* 63 (4):661–663.

Lishak, R. S. 1984. Alarm vocalizations of adult gray squirrels. *Journal of Mammalogy* 65 (4):681–687.

Lybecker, D., B. L. Lamb, and P. D. Ponds. 2002. Public attitudes and knowledge of the black-tailed prairie dog: a common and controversial species. *BioScience*. 52 (7):607–613.

Lyman, C. P., J. S. Willis, A. Malan, and L. C. H. Wang. 1982. *Hibernation and Torpor in Mammals and Birds*. Academic Press, New York.

MacDonald, I. M. V. 1997. Field experiments on duration and precision of grey and red squirrel spatial memory. *Animal Behaviour* 54 (4):879–891.

MacKinnon, K. S. 1978. Stratification and feeding differences among Malayan squirrels. *Malayan Nature Journal* 30 (3/4):593–608.

Manski, D. A., L. W. VanDruff, and V. Flyger. 1981. Activities of gray squirrels and people in a downtown Washington, D.C. park: Management implications. *Transactions of the North American Wildlife and Natural Resources Conference* 46:439–454.

Mateo, J. M. 2003. Kin recognition in ground squirrels and other rodents. *Journal of Mammalogy* 84 (4):1163–1181.

McNulty, F. 1970. *Must They Die? The Strange Case of the Prairie Dog and the Black-Footed Ferret*. Doubleday, New York.

Mercer, J. M., and V. L. Roth. 2003. The effects of Cenozoic global change on squirrel phylogeny. *Science* 299 (5612):1568–1572.

Merriam, C. H. 1884. *The Mammals of the Adirondack Region*. Press of L. S. Foster, New York.

Merriam, C. H. 1910. The California ground squirrel. USDA Bureau of Biological Survey Circular 76:1–15.

Merritt, J. F., D. A. Zegers, and L. R. Rose. 2001. Seasonal thermogenesis of southern flying squirrels (*Glaucomys volans*). *Journal of Mammalogy* 82 (1):51–64.

Michener, G. R. 1983. Kin identification, matriarchies, and the evolution of sociality in ground-dwelling sciurids. In *Advances in the Study of Mammalian Behavior*, ed. J. F. Eisenberg and D. G. Kleiman, 528–572. American Society of Mammalogists, Lawrence, KS.

Moore, J. 1959. Relationships among the living squirrels of the Sciurinae. *Bulletin of the American Museum of Natural History* 118:159–206.

Moro, M. H., J. T. Horman, H. R. Fischman, J. K. Grigor, and E. Israel. 1991. The epidemiology of rodent and lagomorph rabies in Maryland, 1981 to 1986. *Journal of Wildlife Diseases* 27 (3):452–456.

Murie, J. O. 1995. Mating behavior of Columbian ground squirrels. 1. Multiple mating by females and multiple paternity. *Canadian Journal of Zoology* 73:1819–1826.

Murie, J. O., and G. R. Michener. 1984. *The Biology of Ground-Dwelling Squirrels*. University of Nebraska Press, Lincoln.

Nagy, K. A. 1994. Seasonal water, energy and food use by free-living, arid-habitat mammals. *Australian Journal of Zoology* 42:55–63.

Neuhaus, P. 2003. Parasite removal and its impact on litter size and body condition in Columbian ground squirrels (*Spermophilus columbianus*). *Proceedings of the Royal Society of London Series B* (Supplement) 270:S213–S215.

Nishida, A. T., S. Uehara, and R. Nyundo. 1979. Predatory behavior among wild chimpanzees of the Mahale Mountains. *Primates* 20 (1):1–20.

Nixon, C. M., L. P. Hansen, and S. P. Havera. 1991. Growth patterns of fox squirrels in East-central Illinois. *American Midland Naturalist* 125 (1):168–172.

Nowak, R. M. 1999. *Walker's Mammals of the World*. Johns Hopkins University Press, Baltimore.

Nowicki, S., and K. B. Armitage. 1979. Behavior of juvenile yellow-bellied marmots: play and social integration. *Zeitschrift für Tierpsychologie* 51 (1):85–105.

Nunes, S., E. M. Muecke, J. A. Anthony, and A. S. Batterbee. 1999. Endocrine and energetic mediation of play behavior in free-living Belding's ground squirrels. *Hormones and Behavior* 36:153–165.

Nunes, S., E. M. Muecke, L. T. Lancaster, N. A. Miller, M. A. Mueller, J. Muelhaus, and L. Castro. 2004. Functions and consequences of play behaviour in juvenile Belding's ground squirrels. *Animal Behaviour* 68:27–37.

Nunes, S., E. M. Muecke, Z. Sanchez, R. R. Hoffmeier, and L. T. Lancaster. 2004. Play behavior and motor development in juvenile Belding's ground squirrels (*Spermophilus beldingi*). *Behavioral Ecology and Sociobiology* 56:97–105.

Ognev, S. I. 1963. *Mammals of the U.S.S.R. and Adjacent Countries*, vol. 5, *Rodents*. Israel Program for Scientific Translation, Jerusalem.

Oli, M. K., I. R. Taylor, and M. E. Rogers. 1993. Diet of the snow leopard (*Panthera uncia*) in the Annapurna Conservation Area, Nepal. *Journal of Zoology (London)* 231 (3):365–370.

O'Shea, T. J. 1976. Home range, social behavior, and dominance relationships in the African unstriped ground squirrel, *Xerus rutilus*. *Journal of Mammalogy* 57 (3):450–460.

Owings, D. H., R. G. Coss, D. McKernon, M. P. Rowe, and P. C. Arrowood. 2001.

Snake-directed antipredator behavior of rock squirrels (*Spermophilus variegatus*): Population differences and snake-species discrimination. *Behaviour* 138:575–595.

Owings, D. H., A. S. Rundus, and M. P. Rowe. 2002. The rattling sound of rattlesnakes (*Crotalus viridis*) as a communicative resource for ground squirrels (*Spermophilus beecheyi*) and burrowing owls (*Athene cunicularia*). *Journal of Comparative Psychology* 116 (2):197–205.

Ozeki, S. I., K. Wakamatsu, and T. Hirobe. 1995. Chemical characterization of hair melanins in various coat-color mutants of mice. *Journal of Investigative Dermatology* 105:361–366.

Paschoal, M., and M. Galetti. 1995. Seasonal food use by the Neotropical squirrel *Sciurus ingrami* in Southeastern Brazil. *Biotropica* 27 (2):268–273.

Patton, J. L., M. N. F. Da Silva, and J. R. Malcolm. 2000. Mammals of the Rio Juruá and the evolutionary and ecological diversification of Amazonia. *Bulletin of the American Museum of Natural History* 244:1–306.

Pauli, J. N. 2005. Evidence for long-distance swimming capabilities in red squirrels, *Tamiasciurus hudsonicus*. *Northeastern Naturalist* 12 (2):245–248.

Payne, J. B. 1979. Synecology of Malayan tree squirrels with special reference to the genus *Ratufa*. Ph.D. thesis. University of Cambridge, Cambridge.

Payne, J. B. 1980. Competitors. In *Malayan Forest Primates: Ten Year's Study in Tropical Rain Forest*, ed. D. J. Chivers, 261–278. Plenum Press, New York.

Pocock, R. I. 1923. On the classification of the Sciuridae. *Proceedings of the Zoological Society of London* 1:209–246.

Poran, N. S., R. G. Coss, and E. Benjamini. 1987. Resistance of California ground squirrels (*Spermophilus beecheyi*) to the venom of the Northern Pacific rattlesnake (*Crotalus viridis oreganus*): a study of adaptive variation. *Toxicon* 25 (7):767–777.

Potter, B. 1903. *The Tale of Squirrel Nutkin*. Frederick Warne & Co., London.

Prosser, C. L., ed. 1991. *Comparative Animal Physiology*, 4th ed. *Environmental and Metabolic Animal Physiology*. Wiley Liss, Inc., New York.

Rayor, L. S., and K. B. Armitage. 1991. Social behavior and space-use of young of ground-dwelling squirrel species with different levels of sociality. *Ethology, Ecology, and Evolution* 3:185–205.

Réale, D., A. G. McAdam, S. Boutin, and D. Berteaux. 2003. Genetic and plastic responses of a northern mammal to climate change. *Proceedings of the Royal Society of London Series B* 270: 591–596.

Salten, F. 1929. *Bambi*. Grosset & Dunlap, New York.

Schmeltz, L. L., and J. O. Whitaker Jr. 1977. Use of woodchuck burrows by woodchucks and other mammals. *Transactions of the Kentucky Academy of Science* 38 (1–2):79–82.

Scholey, K. 1986. The climbing and gliding locomotion of the giant red flying squirrel *Petaurista petaurista* (Sciuridae). *BIONA-Report* 5:187–204.

Schulte-Hostedde, A. I., and J. S. Millar. 2002. Effects of body size and mass on running speed of male yellow-pine chipmunks (*Tamias amoenus*). *Canadian Journal of Zoology* 80:1584–1587.

Searle, A. G. 1968. *Comparative Genetics of Coat Colour in Mammals*. Logos Press, London.

Seebeck, J. H. 1989. *Sciuridae*. In *Fauna of Australia*, ed. D. W. Walton, Vol. 1B, Australian Government Publishing Service, Canberra.

Semennov, Y., R. Ramousse, M. Le Berre, and Y. Tutukarov. 2001. Impact of the black-capped marmot (*Marmota camtschatica bungei*) on floristic diversity of arctic tundra in Northern Siberia. *Arctic Antarctic and Alpine Research* 33 (2):204–210.

Sherman, P. W. 1977. Nepotism and the evolution of alarm calls. *Science* 197 (4310):1246–1253.

Silvers, W. K. 1979. *The Coat Colors of Mice: A Model for Mammalian Gene Action and Interaction*. Springer-Verlag, New York.

Sludskiĭ, A. A. 1969. *Mammals of Kazakhstan*, vol. 1. Alma-Alta, Kazakhstan.

Smallwood, P. D., and W. D. Peters. 1986. Gray squirrel food preferences: The effects of tannin and fat concentration. *Ecology* 67 (1):168–174.

Smallwood, P. D., M. A. Steele, and S. H. Faeth. 2001. The ultimate basis of the caching preferences of rodents and the oak-dispersal syndrome: tannins, insects, and seed germination. *American Zoologist* 41:840–851.

Smith, C. C. 1970. The coevolution of pine squirrels (*Tamiasciurus*) and conifers. *Ecological Monographs* 40 (3):349–371.

Smith, C. C. 1981. The indivisible niche of *Tamiasciurus*: an example of nonpartitioning of resources. *Ecological Monographs* 51 (3):343–363.

Smith, C. C. 1995. The niche of the diurnal tree squirrel. In *Storm over a Mountain Island: Conservation Biology and the Mt. Graham Affair*, ed. C. A. Istock and R. S. Hoffmann, 209–225. University of Arizona Press, Tucson.

Sobelman, S. S. 1985. The economics of wild resource use in Shishmaref, Alaska. Alaska Department of Fish and Game Technical Paper No. 112.

Stapanian, M. A., and C. C. Smith. 1978. A model for seed scatterhoarding: coevolution of fox squirrels and black walnuts. *Ecology* 59 (5):884–896.

Stapp, P. 1998. A reevaluation of the role of prairie dogs in Great Plains grasslands. *Conservation Biology* 12 (6):1253–1259.

Steele, M. A., L. Z. Hadj-Chikh, and J. Hazeltine. 1996. Caching and feeding decisions by *Sciurus carolinensis:* responses to weevil-infested acorns. *Journal of Mammalogy* 77 (2):305–314.

Steele, M. A., T. Knowles, K. Bridle, and E. L. Simms. 1993. Tannins and partial consumption of acorns: implications for dispersal of oaks by seed predators. *American Midland Naturalist* 130 (2):229–238.

Steele, M. A., and J. L. Koproski. 2001. *North American Tree Squirrels*. Smithsonian Institution Press, Washington, DC.

Steele, M. A., P. D. Smallwood, A. Spunar, and E. Nelsen. 2001. The proximate basis of the oak dispersal syndrome: detection of seed dormancy by rodents. *American Zoologist* 41:852–864.

Steele, M. A., G. Turner, P. D. Smallwood, J. O. Wolff, and J. Radillo. 2001. Cache management by small mammals: experimental evidence for the significance of acorn-embryo excision. *Journal of Mammalogy* 82 (1):35–42.

Steppan, S. J., M. R. Akhverdvan, E. A. Lyapunova, D. G. Fraser, N. N. Voronstov, R. S. Hoffmann, and M. J. Braun. 1999. Molecular phylogeny of the marmots

(Rodentia: Sciuridae): Tests of evolutionary and biogeographic hypotheses. *Systematic Biology* 48 (4):715–734.

Steppan, S. J., B. L. Storz, and R. S. Hoffmann. 2004. Nuclear DNA phylogeny of the squirrels (Mammalia: Rodentia) and the evolution of arboreality from c-myc and RAG1. *Molecular Phylogenetics and Evolution* 30 (3):703–719.

Sullivan, T. P., and W. Klenner. 1993. Influence of diversionary food on red squirrel populations and damage to crop trees in young lodgepole pine forest. *Ecological Applications* 3 (4):708–718.

Swaisgood, R. R., M. P. Rowe, and D. H. Owings. 1999. Assessment of rattlesnake dangerousness by California ground squirrels: exploitation of cues from rattling sounds. *Animal Behaviour* 57 (6):1303–1310.

Swihart, R. K., and P. M. Picone. 1994. Damage to apple trees associated with woodchuck burrows in orchards. *Journal of Wildlife Management* 58 (2):357–360.

Tamura, N., and H. Yong. 1993. Vocalizations in response to predators in three species of Malaysian *Callosciurus* (Sciuridae). *Journal of Mammalogy* 74 (3):703–714.

Tangney, D., and H. McCann. 1996. Red squirrel swimming in Lower Lough Erne. *Irish Naturalist Journal* 25 (8):304.

Tennant, B. C., and J. L. Gerin. 2001. The woodchuck model of hepatitis B virus infection. *ILAR Journal* 42 (2):89–102.

Thomas-Lester, A. 1990. Kamikaze squirrels? Nope, just more of them running around. *Washington Post*, 9 October, B1–B4.

Thorington, R. W., Jr. 1966. The biology of rodent tails: a study of form and function. Ph.D. thesis. Arctic Aeromedical Laboratory, Fort Wainwright, AL.

Thorington, R. W., Jr., and K. Darrow. 1996. Jaw muscles of Old World squirrels. *Journal of Morphology* 230:145–165.

Thorington, R. W., Jr., and K. Darrow. 2000. Anatomy of the squirrel wrist: bones, ligaments and muscles. *Journal of Morphology* 246:85–102.

Thorington, R. W., Jr., K. Darrow, and C. G. Anderson. 1998. Wing tip anatomy and aerodynamics in flying squirrels. *Journal of Mammalogy* 79 (1):245–250.

Thorington, R. W., Jr., and L. R. Heaney. 1981. Body proportions and gliding adaptations of flying squirrels (Petauristinae). *Journal of Mammalogy* 62 (1):101–114.

Thorington, R. W., Jr., and R. S. Hoffman. 2005. Family Sciuridae. In *Mammal Species of the World: A Taxonomic and Geographic Reference*, ed. D. E. Wilson and D. M. Reeder, vol. 2, 754–818, 3rd ed. Johns Hopkins University Press, Baltimore.

Thorington, R. W., Jr., C. E. Schennum, L. A. Pappas, and D. Pitassy. 2005. The difficulties of identifying flying squirrels (Sciuridae: Pteromyini) in the fossil record. *Journal of Vertebrate Paleontology* 25 (4):950–961.

Tompkins, D. M., A. W. Sainsbury, P. Nettleton, D. Buxton, and J. Gurnell. 2002. Parapoxvirus causes a deleterious disease in red squirrels associated with UK population declines. *Proceedings of the Royal Society of London Series B* 269:529–533.

Topsell, E. 1967. *The History of Four-Footed Beasts and Serpents and Insects*, vol. 1. Da Capo Press, New York.

Vander Wall, S. B. 1990. *Food Hoarding in Animals*. University of Chicago Press, Chicago.

Vander Wall, S. B. 1991. Mechanisms of cache recovery by yellow pine chipmunks. *Animal Behaviour* 41 (5):851–863.

Vander Wall, S. B. 2000. The influence of environmental conditions on cache recovery and cache pilferage by yellow pine chipmunks (*Tamias amoenus*) and deer mice (*Peromyscus maniculatus*). *Behavioral Ecology* 11 (5):544–549.

Vander Wall, S. B. 2001. The evolutionary ecology of nut dispersal. *Botanical Review* 67 (1):74–117.

Vander Wall, S. B., and S. H. Jenkins. 2003. Reciprocal pilferage and the evolution of food hoarding behavior. *Behavioral Ecology* 14 (5):656–667.

van Heerden, J., and J. Dauth. 1987. Aspects of adaptation to an arid environment in free-living ground squirrels *Xerus inauris*. *Journal of Arid Environments* 13:83–89.

Vernes, K. 2001. Gliding performance of the northern flying squirrel (*Glaucomys sabrinus*) in mature mixed forest of eastern Canada. *Journal of Mammalogy* 82 (4):1026–1033.

Vesey, T. 1984. The squirrels who can zap Thanksgiving. *Washington Post* 21 November, B1.

Vianey-Liaud, M. 1974. *Palaeosciurus goti* nov. sp., écureuil terrestre de l'Oligocène moyen du Quercy. Données nouvelles sur l'apparition des Sciuridés en Europe. *Annales de Paléontologie (Vertébrés)* 60 (1):103–122.

Viljoen, S. 1977. Behaviour of the bush squirrel, *Paraxerus cepapi cepapi* (A. Smith, 1836). *Mammalia* 42 (2):119–166.

Walls, G. L. 1942. *The Vertebrate Eye and its Adaptive Radiation*. Cranbook Institute of Science, Michigan.

Walsberg, G. E. 1988. Consequences of skin color and fur properties for solar heat gain and UV irradiance in two mammals. *Journal of Comparative Physiology B Biochemical Systemic and Environmental Physiology* 15 8(2):213–222.

Walsberg, G. E., T. Weaver, and B. O. Wolf. 1997. Seasonal adjustment of solar heat gain independent of coat coloration in a desert mammal. *Physiological Zoology* 70 (2):150–157.

Waterman, J. M. 1988. Social play in free-ranging Columbian ground squirrels, *Spermophilus columbianus*. *Ethology* 77 (3):225–236.

Waterman, J. M. 1995. The social organization of the Cape ground squirrel (*Xerus inauris*; Rodentia: Sciuridae). *Ethology* 101:130–147.

Waterman, J. M. 1996. Reproductive biology of a tropical non-hibernating ground squirrel. *Journal of Mammalogy* 77 (1):134–146.

Waterman, J. M. 1997. Why do male Cape ground squirrels live in groups? *Animal Behaviour* 53:809–817.

Waterman, J. M. 1998. Mating tactics of male Cape ground squirrels, *Xerus inauris*: consequences of year-around breeding. *Animal Behaviour* 56:459–466.

Wauters, L. A., G. Tosi, and J. Gurnell. 2002. Interspecific competition in tree squirrels: do introduced grey squirrels (*Sciurus carolinensis*) deplete tree seeds hoarded by red squirrels (*Sciurus vulgaris*)? *Behavioral Ecology and Sociobiology* 51:360–367.

Wauters, L. A., M. Zaninetti, G. Tosi, and S. Bertolino. 2004. Is coat-colour polymorphism in Eurasian red squirrels (*Sciurus vulgaris* L.) adaptive? *Mammalia* 68 (1):37–48.

Weeks, H. P., and C. M. Kirkpatrick. 1978. Salt preferences and sodium drive phenology in fox squirrels and woodchucks. *Journal of Mammalogy* 59 (3):531–542.

Weigl, P. D. 1978. Resource overlap, interspecific interactions and the distribution of the flying squirrels, *Glaucomys volans* and *G. sabrinus*. *American Midland Naturalist* 100 (1):83–96.

Weigl, P. D., and E. V. Hanson. 1980. Observational learning and the feeding behavior of the red squirrel *Tamiasciurus hudsonicus:* the ontogeny of optimization. *Ecology* 61 (2):213–218.

Weigl, P. D., L. J. Sherman, and W. J. Grundman. 1983. Body size, food size and the ecology of tree squirrels. Paper presented at the 63rd annual meeting of the American Society of Mammalogists, University of Florida, Gainesville.

Wetzel, E. J., and P. D. Weigl. 1994. Ecological implications for flying squirrels (*Glaucomys spp.*) of effects of temperature on the in vitro development and behavior of *Strongyloides robustus*. *American Midland Naturalist* 131 (1):43–54.

Whisson, D. A., S. B. Orloff, and D. L. Lancaster. 1999. Alfalfa yield loss from Belding's ground squirrels in northeastern California. *Wildlife Society Bulletin* 27 (1):178–183.

Whitten, J. E. J., and A. J. Whitten. 1987. Analysis of bark eating in a tropical squirrel. *Biotropica* 19 (2):107–115.

Widen, P. 1987. Goshawk predation during winter, spring and summer in a boreal forest area of central Sweden. *Holarctic Ecology* 10 (2):104–109.

Wilson, D. R., and J. F. Hare. 2004. Ground squirrel uses ultrasonic alarms. *Nature* 430:523.

Wishner, L. 1982. *Eastern Chipmunks: Secrets of their Solitary Lives.* Smithsonian Institution Press, Washington, DC.

Zahler, P. 1996. Rediscovery of the woolly flying squirrel. *Journal of Mammalogy* 77 (1):54–57.

Zahler, P. 2001. The woolly flying squirrel and gliding: does size matter? *Acta Theriologica* 46 (4):429–435.

Zahler, P., and M. Khan. 2003. Evidence for dietary specialization on pine needles by the woolly flying squirrel (*Eupetaurus cinereus*). *Journal of Mammalogy* 84 (2):480–486.

Zahler, P., and C. A. Woods. 1997. The status of the woolly flying squirrel *Eupetaurus cinereus* in northern Pakistan. In *Biodiversity of Pakistan*, ed. S. A. Mufti, C. A. Woods, and S. A. Hasan, 485–514. Florida Museum of Natural History, Gainesville.

Index

Adler, Bill, 127
Aengus, 147
Aesculus, 122
agouti, 41–42, 44
alarm call, 56–57, 60, 62–66, 142
albinism, 42–43, 142
albino, 38, 42, 44
alleles, 41–42
Allen, Durward, 152
Alpinia kwangsiensis, 105
altricial, 98
altruistic, 85
Amanita muscaria, 104
Amanita phalloides, 104
Ammospermophilus, 4, 27, 77, 99; *harrisii*, 99; *leucurus*, 20, 77
Ando, Motokazu, 36
antelope squirrel, 4, 7, 27, 70
ants, 104
Armitage, Kenneth, 49–50, 82, 153
army ants, 59
astragalus, 3, 19
Austad, Steven, 101

bacteria, 86–87
baculum, 16, 18
ball-chain collars, 156
Bambi, 148
Barash, David, 61–62, 153
Barbera, Hanna, 144
bark gleaner, 16, 74, 105
bark stripping, 124
Barkalow, Fred, 55, 96, 100, 114, 126, 152
Barro Colorado Island, 72
behavior, 49–52, 55–56, 60, 63, 66–67, 78, 85, 87, 92–93, 104–6, 111–12, 114, 117, 142, 145, 148, 150, 155
Benkman, Craig, 89
Bezuidenhout, Abraham, 154
blind spot, 61
Blumstein, Daniel, 49, 64
body fat, 56, 77, 80, 97, 103, 108, 110
Borges, Renee, 153
bot flies, 86
Boutin, Stan, 135
brain size, 154
bruchid beetle, 121
Bryant, Andrew, 133
Bryant, William Cullen, 146
Bumiller, Elisabeth, 124
Burroughs, John, 29, 148, 151

bush squirrel, 7, 14, 19, 20, 73, 104, 155; green, 98

calcaneus, 3, 19
Callosciurinae, 13–19
Callosciurus, 17, 41–43, 47, 58, 64, 73–74, 125, 138, 155; *caniceps*, 74; *erythraeus*, 138; *ferrugineus*, 41; *finlaysonii*, 41–42, 47; *melanogaster*, 58; *nigrovittatus*, 74; *notatus*, 73; *prevostii*, 17, 42, 47, 73
camouflage, 41, 58–59, 66
Candlemas, 143
cannibalism, 105
Carya, 122
Castanea, 122
Castanopsis, 122
cecum, 106
Centers for Disease Control and Prevention, 130
Chambers, Whittaker, 148
cheek pouch, 20, 27–28
Cherokee, 142
chickaree, 40
chimpanzee, 84
Chip and Dale, 144
chipmunk, 4, 7, 9, 14, 16–17, 19–20, 26–28, 30, 41–44, 50, 59, 63–64, 70, 72, 80–81, 86, 93–94, 103, 108, 110; eastern, 9, 28, 41–42, 63–64, 72, 153; least, xiii, 16; Siberian, 138; yellow pine, 113
cicada, 103
Cinderella, 143
clade, 11
climate change, 134, 138
coat color, 38, 40–41, 43–47, 117, 155
coat patterns, 43
cocoa, 125
color gene, 42
color vision, 26–27
cone: of the eyes, 27; of the teeth, 2; of trees, 88–89, 102, 104, 108–10, 136, 150
Corylus, 122
Coss, Richard, 67
coterie, 70, 85–86, 94–95
countercurrent, 6
counter-shading, 43, 58
cowpox, 87
Crosby, Ernest, 146–47
cryptomeria, xiii
Cynomys, xiv, 4, 20; *leucurus*, 20; *ludovicianus*, xiv

Death Angel mushroom, 104
deltoid muscle, 4, 20
Devenport, Jill, 112
diastema, 2
Douglas fir, 102
Douglassciurus jeffersoni, 23

eagle, golden, 120
ectoparasites, 49, 86
Eliot, George, 150
Elliot, Philip, 89
Elliott, Lang, 64
Emerson, Ralph Waldo, 144–45, 149
Emmons, Louise, 16, 64, 73, 75, 98, 106, 108, 153, 155
Emry, Robert, 23
endorhyzal fungi, 122
endotherms, 79
Eocene, 13, 23
Epixerus, 20, 73, 153; *ebii*, 73
eumelanin, 41, 42
Eupetaurus cinereus, 9, 17, 31, 37, 105, 154
Evans, Howard, 154
Exilisciurus, 17, 25, 105; *whiteheadi*, 17
eye lens, 26
eye rings, 44
eyesight, 42–43, 118

Fagus, 122
Fairbairn, Douglas, 149
fatty acids, 80, 104
feeding, 102
ferret, black-footed, 84, 120
fetid marigold, 121
Field Museum, xiii
figs, 104
fleas, 86, 87, 116
Fly Amanita mushroom, 104
Flyger, Vagn, 28, 119, 129
flying squirrel: complex-toothed, 9, 154;
    Japanese, 118; Japanese giant, xiii, 36–37,
    118, 154; pygmy, 17, 32, 34, 134–35, 154;
    Siberian, 118; woolly, 17, 37, 69, 105, 154
follicle, 41
foods, 102
fovea centralis, 27
fox, 84, 121, 129–30; swift, 120
Frazer, Sir James, 142
frogs, 104
Funambulus, 16–18, 43, 122, 138–39, 141, 155;
    palmarum, 141, 155; *pennantii*, 18
fungal spores, 122
fungi, 104–5, 122, 124
Funisciurus, 20, 21, 43, 73, 87, 153, 155;
    anerythrus, 73; *isabella*, 73; *lemniscatus*, 73;
    pyrropus, 73, 155

genes, 12, 19, 40–43, 84–85
genetic, 21, 41, 44

Gerhardt, Fritz, 109, 152
Gesner, Conrad, 148
gestation, 59, 95, 98
giant, 32
glacial refugia, 76
Glanz, Bill, 72
Glaucomys, 9, 10, 34, 36, 72, 81, 82, 84, 102,
    104, 118–20, 154; *sabrinus*, 9, 36, 72, 82, 104,
    118, 120, 154; *volans*, 10, 34, 36, 72, 81, 84,
    102, 118–19, 154
gliding, 31
global warming, 134–36
Golden Bough, 142
goshawk, 120
Great Plains ecosystem, 120
ground squirrel: African striped, 95, 125;
    African unstriped, 95; Arctic, 120, 122, 123;
    California, 64, 66–67, 99, 103; Columbian,
    55, 97, 153; golden mantled, 95; long-nosed,
    17, 74; Mohave, 77–78; round-tailed, 77–78;
    South African, 71, 95, 97, 153; thirteen-lined,
    20, 41, 44, 50, 82, 97; three-striped, 74
Groundhog Day, 142
Gurnell, John, 152

Hadj-Chikh, Leila, 111
hair dye, 156
Hamilton, William, 28
Handley, Charles, 72
Hanson, Elinor, 107
Harvard Law, 9
hawk, 26, 59, 64, 150; ferruginous, 120
Haynie, Michelle, 97
Heaney, Larry, xiii
hearing, 61, 64, 115, 118, 150
heart rate, 25–26, 77, 80, 82
Heliosciurus, 20, 73, 87, 105, 153, 155;
    gambianus, 155; *rufobrachium*, 73, 105, 155
hemlock, western, 102
hepatitis, 88, 122
hickory nut, 106
Hobbit, 142
Hodos, William, 52
Holmes, Donna, 101
home range, 49, 69, 70–71, 88, 93, 101, 108,
    112, 117
Hood, Thomas, 145
Hoogland, John, 50, 52, 62, 85, 97, 100, 153
Hopi, 139
horned larks, 120
Hrdlička, Aleš, 38
hunting, 151

Imaizumi, Yoshimori, xiii
incisors, 1–3, 27, 59, 89–90, 106, 115, 129
infanticide, 50, 84–85, 123
Inouye, David, 135
insect larvae, 104, 112
insects, 73–75, 79, 102, 104, 105, 112, 121, 124

interpluvial period, 76
introduced species, 137
Iroquois, 141

Jacobs, Lucia, 112
Jones, Clyde, 115
Juglans, 122

Kalahari, 76
Kalevala, 142
Kazan hairs, 140
Keats, John, 146
Kentucky long rifle, 138
keystone species, 120
Kiltie, Richard, 59
kin recognition, 52, 60
Kinabalu, 72
Kipling, Rudyard, 146–47
Kirkpatrick, Charles, 103
Koprowski, John, 93, 100, 110, 152
Kotiliar, Natasha, 120

lactation, 98
Lafayette Park, 126, 128
larder-hoarding, 108–10
Lariscus, 43, 74, 155; *insignis*, 74
learning, 54, 107, 118, 123
Leopold, Aldo, 119
lice, 86
Liman, Emily, 112–13
lions, 85, 130
Lithocarpus, 122
litter size, 50, 86, 96, 101
lodgepole pine, 89
Lompat, Kuala, 73–74
Lyme disease, 130

MacDonald, Isabel, 113
MacKinnon, Kathleen, 73
maize, 125
marmot, 131, 133, 138; Alpine, 64, 138; black-
    capped, 83, 94; Bobak, 80, 94; gray, 25–26;
    Himalayan, 9, 120; Olympic, 48, 55, 60, 62–
    63, 100; Vancouver Island, 41, 64, 134–35;
    yellow-bellied, 25, 153
Marmota, 4, 9, 20, 21, 25, 29, 50, 94, 100, 103,
    108, 122, 130, 134, 135, 138, 139, 143, 153;
    bobak, 94; *camtschatica*, 83, 94; *himalaya*, 9;
    marmota, 138; *monax*, 9, 20, 29, 50, 100, 103,
    108, 122, 130, 143; *olympus*, 100; *sibirica*, 139
Marmotini, 6, 14, 15, 19, 20
mating chase, 52, 57, 92–95
meadow vole, 121
melanin, 41, 42
melanism, 39, 40, 59
melanistic, 38–40, 42, 59
melanoblast, 44
melanocyte, 41
Melville, Herman, 149

memory, 54, 112–13, 142
Menetes, 43, 155
Mercer, John, 13–14
Meredith, George, 147
Merriam, C. Hart, 29
metabolic rates, 79, 105
Michener, Gail, 48
Microsciurus, 72–73, 155; *alfari*, 72
migration, 44, 71
Miocene, 14, 67
Mirkwood, 142
mites, 86, 116
Mohave Desert, 77
monkeypox, 87, 114, 139
Moore, Joseph, 16, 18, 19, 21, 154
morph, 38, 46
mountain plovers, 120
mouse: deer, 120; grasshopper, 120; house, 121;
    meadow jumping, 121; white-footed, 121
Murie, Jan, 97
mushrooms, 104, 112
Mustela nigripes, 120
Muul, Illar, 37, 134
Myosciurus, 20–22, 25–26, 73, 105; *pumilio*, 22,
    25–26, 73

Nannosciurus, 17; *melanotis*, 17
National Zoological Park, 31, 38–39, 42
Navajo, 139
nepotism, 60, 65
nesting, 51, 69, 95–96, 115, 129
Neuhaus, Peter, 86
neural crest cell, 44
Nidhogg, 142
nightshade, black, 121
nonagouti, 41–42
nuclear genes, 12
Nunes, Scott, 56

obligate hibernator, 80
Oligocene, 13–14, 23–24
Olney, Illinois, 42
Owings, Donald, 67
owl, burrowing, 120

Paleosciurus goti, 24
palm nut, 104, 106, 108, 125, 155
palm squirrel, Indian, 43–44, 138, 141, 155
Panama, 72, 74, 155
Panama land bridge, 74
parapox, 87, 137
Paraxerus, 20–21, 41, 73, 98, 104, 153, 155;
    cepapi, 98; *poensis*, 73, 98, 104; *vexillarius*,
    41
Pauli, Jonathan, 29
Payne, Junaidi (formerly John), 73, 153
perianal scent gland, 57
perineum, 57
Perri, 148

Petaurillus, 17, 32, 134–35, 154; *kinlochii*, 32, 134–35, 154
Petaurista, 31–32, 36–37, 118, 154; *leucogenys*, 36, 118, 154
phaeomelanin, 41–42
phalange, 2–3
pigweed, 121
pine needle, 105, 154
Pinus koraiensis, 122
pisiform, 18, 33, 35
pit tags, 156
plague, 50, 86–87, 116, 120, 125
play, 5, 49, 55–57, 110–11, 120–22, 141–42, 144, 146
Pleistocene, 76
Pocock, Reginald, 14, 16, 21
pollination, 122
Ponderosa pine, 89, 102
Poran, Naomie, 67
Potter, Beatrix, 148
prairie dog, 7, 19, 20, 27, 48, 50–53, 55, 57, 60–62, 64–65, 70, 84–85, 87–88, 94, 97, 100, 114, 116, 120–21, 123, 125, 139; black-tailed, xiv, 21, 52, 80, 95, 125
predator, 59–61, 63–65, 84, 91, 140
premolar, 2
Procyon lotor, 130
pronation, 33–34
Protoxerini, 14–15, 19–20
Protoxerus, 20–21, 26, 73, 153; *stangeri*, 21, 26, 73
Pteromyini, 14–15, 17
Pteromys: *momonga*, 118; *volans*, 102, 118, 154
Punxsutawney Phil, 143

Quercus, 122

rabbit, cottontail, 121
rabies, 130
raccoon, 121, 129–30
radio collar, 156
radius, 24, 33–34, 37
Ramayana, 141, 144
Ratatosk, 142
Rattus norvegicus, 87
Ratufa, 16, 25, 30, 73, 106, 153; *affinis*, 73, 106, 153; *bicolor*, 47, 73, 153; *indica*, 16, 30, 153
Ratufinae, 13–15, 17, 19
Ravana, 141
Reale, Denis, 135
red squirrel, 40, 144
retina, 27
Rheithrosciurus, 3, 17
Rhinosciurus, 17, 74, 105, 155; *laticaudatus*, 17, 74, 105
rock squirrel, 9, 27, 86
Rodentia, 13, 15
rods, 27
Roosevelt, Theodore, 124, 151

rope squirrel, 19–20, 43, 73, 155; fire-footed, 155
Roth, V. Louise, 13–14
roundworms, 86

Sabrosky, Curt, 140
Sacamena hairs, 140
Salten, Felix, 148
scapholunate, 18, 33
scarlet mallow, 121
Sciuridae, 1, 13–15
Sciurillinae, 13–15
Sciurillus, 10, 18, 73, 105
Sciurinae, 13–17, 19
Sciurini, 14–19
sciuromorphy, 2
Sciurus, 2, 9, 11, 18, 38, 40, 46, 71–73, 90, 92, 102–6, 108, 111, 125, 137–39, 142–43, 148, 152, 154–55; *aberti*, 102, 104; *aestuans*, 73, 106, 108; *aureogaster*, 111; *carolinensis*, 2, 9, 11, 38, 72, 102, 108, 137; *gilvigularis*, 72; *granatensis*, 40, 72, 92, 125, 154; *igniventris*, 72; *lis*, 104; *variegatoides*, 72, 154; *vulgaris*, 40, 46, 63, 71, 90, 102, 108, 137, 139, 142–43, 148, 152
Secret Squirrel, 144
Seneca, 141
Sherman, Paul, 65
Shiraishi, Satoshi, 36
Shorten, Monica, 55, 152
shrew: masked, 121; short-tailed, 121
silver, 43, 123
Sita, 141
skull, 1–14, 16, 19, 23, 26–27, 32
smallpox, 87
Smallwood, Peter, 112
Smith, Chris, 58, 94, 103
snake, 65
sodium, 103
Sonoran Desert, 76
Spermophilopsis leptodactylus, 9, 19
Spermophilus, 4, 6, 9–11, 20, 22, 27, 43–45, 50, 62, 67, 77–79, 96–97, 99, 112, 122; *beecheyi*, 99; *beldingi*, 62, 96–97, 99; *mexicanus*, 67; *mohavensis*, 77; *parryii*, 122; *richardsonii*, 4, 11, 97; *tereticaudus*, 77; *townsendii*, 78; *tridecemlineatus*, 20, 44, 50, 97, 112; *variegatus*, 9, 67, 79
spots, 44, 60–61, 63, 109
spotted owl, 84, 120
squirrel: Arizona gray, 51; black, 38–40; black-striped, 74; Delmarva fox, 46; eastern fox, 46; eastern gray, 28, 38, 41–43, 45, 46, 51–52, 56–58, 63, 72, 87, 101, 108, 111, 113, 126, 137–38; Eurasian red, 28, 41–42, 46, 51, 71, 87, 90, 104, 108, 120, 137–140, 148, 153; gray-bellied, 74; Guianan, 73, 106, 108; horse-tailed, 74; Japanese, 104; Mexican fox, 51; Mexican gray, 111; Mount Graham red,

136; red-tailed, 40; variegated, 154. *See also* bush squirrel; flying squirrel; ground squirrel; palm squirrel; rope squirrel; tree squirrel
Squirrel Nutkin, 148
squirrel skins, 123
squirrel-hair brushes, 139
Stafford, Brian, 37
Steele, Michael, 110–12, 152
Steppan, Scott, 13–14
stripes, 19, 22, 43–45, 141–42
Strongyloides robustus, 136
styliform cartilage, 3, 19
sun squirrel, 87, 93; Gambian, 155; red-legged, 155
Sundasciurus: *hippurus*, 74; *lowii*, 74; *tenuis*, 73
supination, 33–34
swimming, 28–29, 71, 148

Talahoutky hairs, 140
talus, 3
Tamias, 4, 9, 16–17, 20, 28, 43, 45, 50, 70, 72, 113, 138; *amoenus*, 113; *senex*, 9; *sibiricus*, 138; *striatus*, 9, 28, 70, 72; *townsendii*, 9
Tamiasciurus: *douglassii*, 89; *hudsonicus*, 89, 136
Tamiops, 17, 43, 105, 122, 155; *swinhoei*, 105, 155
Tamura, Noriko, 64
tannins, 90–91
tapeworms, 86
teeth, 1, 2, 14, 20, 37, 106, 114, 126, 154
Thayer, Abbott, 58
thermoregulation, 40
Thomas-Lester, Avis, 71
Thoreau, Henry David, 150
ticks, 86
Tokyo, 64
Tolkien, J. R. R., 142
Tompkins, Daniel, 137
Topsell, Edward, 148
torpor, 82
tree shrew, 155
tree squirrel: Prevost's, 14, 17, 47, 73; pygmy, xiii, 7, 13, 25, 32, 105; red-bellied, 138. *See also* squirrel
triquetral, 33

Trogopterus xanthipes, 9, 36
truffles, 104, 122
Tufty, 144
tulip poplar, 96
Tupaia, 155
tyrosinase, 42

Uinta, 64
ulna, 33–34, 37
unsaturated fats, 103

Vander Wall, Stephen, 88, 113, 121
Vernes, Karl, 36
Vianey-Liaud, Monique, 24
vibrissae, 6
Viljoen, San, 79, 153
virus, 86, 87; squirrel fibroma, 87; West Nile, 130
vision, 26–27, 61, 150
vocalization, 57

Walsberg, Glenn, 79
Waterman, Jane, 49, 153
Wauters, Luc, 137
Weeks, Harmon, 103
weevils, 121
Weigl, Peter, 51, 106–7, 136
White House, 124, 128, 140
Wishner, Lawrence, 59
Wolfe, Humbert, 144
woodchuck, 63, 80, 122
Wordsworth, William, 148

Xerinae, 13–15, 19
Xerini, 6, 14, 15, 19
Xerus erythropus, 70, 95, 125; *inauris*, 50, 66, 71, 77, 95, 97, 153, 154; *princeps*, 71; *rutilus*, 19, 95

Yeats, William Butler, 144–46, 148
Yersinia pestis, 86
Yggdrasil, 142

Zahler, Peter, 37, 154

RICHARD W. THORINGTON JR. works at the National Museum of Natural History of the Smithsonian Institution, where he has spent 35 years as a curator in the Division of Mammals overseeing one of the largest collections of squirrel specimens in the world. He has been observing and studying squirrels for more than 50 years. He has published more than 40 academic papers on squirrel anatomy, behavior, ecology, and systematics. His research on squirrels has included anatomical and systematic studies involving the whole family, Sciuridae, as well as reports on fieldwork in Central America, India, and Japan. He has observed squirrels in South America, Southeast Asia, and in many parts of the United States. His current research is focused on the morphology and systematics of a variety of genera and species of Southeast Asian tree squirrels.

KATIE FERRELL worked as a research assistant at the National Museum of Natural History of the Smithsonian Institution, where she assisted Richard W. Thorington Jr. with his studies on Southeast Asian tree squirrels. She earned both her graduate and undergraduate degrees from Ohio University and has worked previously as a journalist for various publications.